D0906988

AN ORIGINAL THEORY OF THE UNIVERSE

HISTORY OF SCIENCE LIBRARY

Editor: Michael A. Hoskin
Lecturer in the History of Science, Cambridge University

THE ENGLISH PARACELSIANS
Allen G. Debus
Professor of the History of Science, University of Chicago

WILLIAM HERSCHEL AND THE CONSTRUCTION OF THE HEAVENS
M. A. Hoskin
Lecturer in the History of Science, Cambridge University

A HISTORY OF THE THEORIES OF RAIN
W. E. Knowles Middleton

THE ORIGINS OF CHEMISTRY
R. P. Multhauf
Director of the Museum of History and Technology at the Smithsonian Institution

THEORIES OF LIGHT FROM DESCARTES TO NEWTON
A. I. Sabra
Reader in the History of the Classical Tradition, The Warburg Institute

MEDICINE IN MEDIEVAL ENGLAND
C. H. Talbot
Research Fellow of the Wellcome Institute of the History of Medicine, London

THE EARTH IN DECAY
A HISTORY OF BRITISH GEOMORPHOLOGY, 1578-1878
G. L. Davies
Lecturer in the Department of Geography, Trinity College, University of Dublin

THE CONFLICT BETWEEN ATOMISM AND CONSERVATION THEORY 1644-1860
W. L. Scott
Professorial Lecturer, School of Business Administration and Department of Mathematics and Science, American University

THE ROAD TO MEDICAL ENLIGHTENMENT 1650-1695
L. S. King
Professorial Lecturer in the History of Medicine, University of Chicago

WILLIAM HARVEY
AND THE CIRCULATION OF THE BLOOD
Gweneth Whitteridge
Lecturer in the History of Medicine, University of Edinburgh

THE SCIENTIFIC ORIGINS OF NATIONAL SOCIALISM
SOCIAL DARWINISM IN ERNST HAECKEL AND THE GERMAN MONIST LEAGUE
Daniel Gasman
Assistant Professor of History, John Jay College, City University of New York

HISTORY OF SCIENCE LIBRARY: PRIMARY SOURCES

VEGETABLE STATICKS
Stephen Hales
Foreword by M. A. Hoskin
Lecturer in the History of Science, Cambridge University

SCIENCE AND EDUCATION IN THE SEVENTEENTH CENTURY
THE WEBSTER-WARD DEBATE
Allen G. Debus
Professor of the History of Science, University of Chicago

FUNDAMENTA MEDICINAE
Friedrich Hoffmann
Translated and introduced by L. S. King
Professorial Lecturer in the History of Medicine, University of Chicago

THE PNEUMATICS OF HERO OF ALEXANDRIA
Introduced by Marie Boas Hall
Reader in History of Science and Technology, Imperial College

Thomas Wright *of Durham*

AN ORIGINAL THEORY
OR NEW HYPOTHESIS
OF THE UNIVERSE

1750

A facsimile reprint together with the first publication of
A THEORY OF THE UNIVERSE
1734

Introduction and Transcription by
MICHAEL A HOSKIN

MACDONALD · LONDON
AND AMERICAN ELSEVIER INC. · NEW YORK

An Original Theory of the Universe first published 1750
This edition first published 1971

Sole distributors for the United States and Dependencies
American Elsevier Publishing Company, Inc.
52 Vanderbilt Avenue
New York N.Y. 10017

Sole distributors for the British Isles and Commonwealth
Macdonald & Co. (Publishers) Ltd
49–50 Poland Street
London W1A 2LG

All remaining areas
Elsevier Publishing Company
P.O. Box 211
Jan van Galenstraat 335
Amsterdam
The Netherlands

Library of Congress Catalog Card Number 70–139573
Standard Book Numbers
British SBN 356 03515 8
American SBN 444 19612 9

Printed in Great Britain by
Balding and Mansell
London and Wisbech.

CONTENTS

LIST OF PLATES APPEARING IN THE INTRODUCTION

INTRODUCTION

TO THE MACDONALD ELSEVIER EDITION

BY MICHAEL A HOSKIN

THOMAS WRIGHT 'of Durham' was born at Byers Green, near Durham City in the north of England, on 22 September 1711, the son of a yeoman and carpenter.[1] His turbulent boyhood and adolescence were in marked contrast to the serene and ordered adult life that was to follow. His early schooling was cut short 'by a very great Impediment of Speach' and he was, he tells us, 'Very wild & much adicted [to] Sport'. At the age of 13 he was apprenticed to a clock and watch maker, and during this time he was 'Very much given to ye Amusement of Drawing, Planning of Maps and Buildings', developing talents that were to stand him in good stead in later life.

Wright stayed with his master four years, by which time the interest in mathematics and astronomy which had been aroused by his childhood teacher, 'a wrgt good Accomptant and an Astronomer', had become a dominating passion. Encouraged by his mother, Wright was spending all his leisure hours studying astronomy, with such total dedication that 'Father, by ill advice, think him mad. Burn all the Books he can get and endevour to prevent Study.' But soon afterwards, in September 1729, all hope of study had to be abandoned when Wright became involved in a scandal and 'Great Disturbances hapn'd and his Master and Mrs not believing as they should was at last forc'd to run forth'.

After much difficulty Wright managed to reach home, and soon was released from his master by Justices of the Peace. He tried to get work, but without success, so he studied navigation, and wrote 'many things for his own Improvement in astronomy—Geography ec [etcetera]'. He also notes, significantly, 'Reflecting almost upon every object, conseive may find Ideas of ye Deaty and Creation'; and, as we shall see, the task of producing an integrated picture of natural and supernatural, of creation and Creator, was to be the dominating theme of all Wright's cosmological works.

Wright now thought of going to sea, but a storm during his first voyage convinced him that his vocation lay on dry land. He therefore set up a mathematical school at Sunderland in Yorkshire, fell in love with a clergyman's daughter and planned a secret marriage; 'but by an Intelligence from ye B'pp's office at Durham, his Marage was Prevented and Miss Lock'd up'. In his despair Wright headed for London and arranged a passage to the West Indies, but 'being found out by a Friend of my Father's, his going is Prevented'. He then took work in London with makers of mathematical instruments, but after a while his long-suffering father sent money for his journey back north, and Wright once more set up school in Sunderland. There, despite efforts on the part of rivals to have him expelled from the town, he taught navigation to seamen with great success.

With the end of the winter and the departure of the seamen Wright found himself in the spring of 1731 with time on his hands, and he began by planning 'a General Representation of Euclid's Elements in one Large Sheet: and the Doctrine of Plain and Spherical Trigonometry all at one View, on an other'. Such sheets, or 'schemes', were already a feature of the work of other popular teachers of science,[2] but in astronomy Wright was to carry the 'all at one view' technique to the limit and beyond. On p. 1 of the present book Wright speaks of 'a Section of the Creation, eighteen Feet long and one broad' and another 'nine Feet long and six broad'; and, in addition to the sheets of more manageable size which he published on the solar eclipse of 1737 and the comet of 1742, we read in his *Journal* that in 1736 he completed 'his invention of the Theory of Existence and Represent ye Hypotheses in a Section of the Creation 16 feet Long' and the following summer made in brass 'a system of ye Planetary Bodies in true Proportion Equal to a Radius of 190 feet'. In 1737 he drew 'a great many Demonstratory Schemes in Astronomy and Project his Physical and Mathematical Elements', and in his *Journal* he comments: 'His invention at this time run to fast for Execution'. But the most successful scheme was still to come: a chart covering 24 square feet and accompanied by a 'clavis' or key in the form of a substantial volume, *Clavis Cœlestis*, published in 1742.[3]

In addition to planning geometrical and trigonometrical sheets, Wright spent part of 1731 computing and compiling an almanac for the following year, but now his inexperience showed itself: he had allowed insufficient time for the printing, and he had no list of subscribers. The following year he had both time and subscribers to spare, but still had much to learn about the trustworthiness of publishers and engravers. The almanacs were late, and his money 'falling short'. But meanwhile his evident abilities and energy were gaining him influential friends. Wright was a man of lively and versatile intelligence, limited in formal education but full of confidence and willing to turn his mind in any profitable and congenial direction. As an author and teacher of navigation he could no doubt scrape a living; but he realised that as a tutor and consultant to the aristocracy he would share a spacious way of life otherwise denied to him. He had begun modestly enough: in 1731, 'Survey Mr. Aseley's estate at Newfield and make a Map of it.' By autumn 1733 the tide was clearly turning in his favour: 'The Rev. Mr. Newcome introduce to ye Earl of Scarbro', who invites him up to London and Promis's his Patroniseation and Recommendation . . . The Earl of Scarboro' Recommends him to ye Lords of ye Admiralty who give their approbation for Publishing his Pannauticon by Subscription . . . The Earl of Scarboro' obtains leave of the King for him to Dedicate it to His Majesty. Procures him his Royal Highness the Prince for a Subscriber and Recommends him to ye Earls of Ila and Pembroke.'

For many years Wright made a comfortable and agreeable living in high society, holding public and private classes, surveying and advising on the estates of the rich, and publishing a number of books and illustrated sheets. Typical are the entries in his *Journal* for spring 1739: 'Project of a Large Horizontal Dial for Lord Viscount Middleton. Go with his

Lordship to his seat in Surry; stay a month with him. Spend his time extremely agreeable. Lady Midleton Lady Charlotte and Lady Mary Capel study ye use of ye Globes.'

Such was Wright's reputation that in 1742 he was offered £300 a year to become professor of navigation at the Imperial Academy at St Petersburg. He preferred to remain in England, but the note in his *Journal* that 'His Proposals of £500 were sent to Russia' suggests that he was not without his price. Four years later Wright went to Ireland for several months to assemble drawings for his most successful book, *Louthiana*,[4] a description of the antiquities of the County of Louth published in 1748 and again in 1758. But a sequel to *Louthiana* and a volume on the antiquities of England were to remain in manuscript.

In the 1750s Wright's thoughts began to turn once more to his birthplace, no doubt (as Edward Hughes suggests)[5] because nearly all his noble patrons had died off. In 1756 he laid the foundation of a house at Byers Green, and six years later he retired there to 'prosicute my Studies'; many of the Wright manuscripts auctioned by Sotheby's on 19 July 1966, and now in Durham University, belong to this period of retirement. He died in 1786.

Apart from *Louthiana* and a work on *Universal Architecture* which appeared in 1755, Wright's published books all relate to astronomy. The earliest, entitled *Clavis Pannautici, or a Key to the Universal Mariner's Magazine*, is a slim booklet of which Wright's own copy survives at Durham University. It was published in London in 1734, a year after Wright's *Journal* remarks, 'Invents and Completes His Pannauticon, the Universal Mariner's Magazine'. The *Pannauticon* itself has been described as 'a treatise on navigation'[6] and the *Clavis* regarded as 'an introduction to this presumably large volume', and in consequence the absence of any copies of the *Pannauticon* has seemed surprising. But the *Pannauticon* is surely not a book at all; just as the *Clavis Cœlestis* was to be 'the explication of a diagram',[7] so the *Clavis Pannautici* is the guide to the use of an instrument. On p.19 of this guide we read: 'The Instrument consists of four Schemes, one fixt, and three moveable, and delineated in the following Manner . . .', and on p.18 Wright speaks of 'large Schemes eighteen inches Diameter, delineated with divers Circles'. Advertisements on two charts which Wright published for the solar eclipse of 18 February

1736/7 provide the necessary identification. They both state that 'Mr Wright has lately publish'd by Subscription the *Perpetual Pannauticon or Universal Mariners Magazine*, being a Mathematical Instrument', and the chart showing the passage of the penumbra over Scotland specifies that the instrument is '18 inches Diam^{r} describing the Lunar Theory and motion of the Tides'.[8]

Wright's first major work was destined to remain in manuscript. It is a beautifully executed folio volume now at Durham University Library, and is dated 1737. The title runs: '*The Universal Vicissitude of Seasons* exhibiting by inspection at one view, the various rising and setting of the Sun to all parts of the World, with the hour and minute of day-break, length of day, night and twilight etc every day in the year. Illustrated with near 200 copper-plate impressions upon folio paper, all original diagrams curiously delineated, besides the decoration of above 500 various representations of the Sun, and other emblematic ornaments; together with proper tables and demonstratory schemes.' The same library also owns another elaborate unpublished manuscript, undated, in which Wright repeatedly divides and subdivides all knowledge, the divisions being invariably into three parts and resulting eventually in 2187 sections. A similar obsession with triads is to be found in other manuscripts of Wright but nowhere with the thoroughness of this *Pansophia*: a reminder, if one were necessary, that we are dealing with no twentieth-century scientist.

The second of Wright's books to reach publication appeared in 1740: *The Use of Globes*, an introduction to the celestial and terrestrial globes of John Senex with whom Wright had long had business associations and who was to die later the same year. This little book was followed in 1742 by the vast chart (of which only two complete copies are known to survive) and the accompanying key or *Clavis Cœlestis*. The chief merit of this work is, as the author says, his 'Method of treating it, *viz.* clearly and plainly',[9] but there are two features worthy of particular mention. First, Wright points out in words and in drawings how appearances vary with the position of the observer. For example, he tells us (after the style of Christiaan Huygens and more especially James Gregory) that

To *Mercury*, the *Sun* and *Venus* are the only two great Bodies of the Universe. He views *Venus* and all the rest of the Planets, as we do *Saturn*, *Jupiter* and *Mars*;

but *Venus* shines upon him with great Lustre, and 'tis probable, her great Light in opposition to the Sun, serves him instead of a Moon.[10]

Second, and more significantly, Wright does this not only for the planets but also for the stars, which he believes are equal in size and distributed irregularly:

The Stars are looked upon by all modern Astronomers, to be no other than great Globes of Fire like the Sun, promiscuously distributed through the Mundane Space. . . . Their various apparent Magnitudes are intirely the Effect of an unequal Distance as will manifestly appear from the Scheme of their Disposition . . . .[11]

And in the scheme or figure he shows how double stars and even clusters may be optical effects due to the particular location of the observer.

Wright's *Journal* ends in 1746, so our knowledge of his activities in the months immediately preceding the publication in 1750 of his *Original Theory of the Universe* is slight. We do, however, possess a copy of the prospectus of the work.[12] This contains some ten plates and a table of contents, together with a title page:

Now in the Press, and shortly will be publish'd by Subscription, *An Original Theory* . . . . *Half a guinea* to be paid at subscribing, and *Five Shillings* on Delivery of the Book in Sheets, which will be early in the Spring; and if any remain above what are subscribed for, they will be sold at *One Guinea*. Subscriptions are taken in for the Author, at Mr. Pattison's in *Oxenden-Street*, near the *Hay-Market*, and at Mr. Chapelle's, in *Grosvenor-Street*.

Apart from a late work on longitude, the *Original Theory* was to be Wright's last published book in astronomy, and it is his chief claim to a place in the history of the subject. In contrast to the concisely-factual approach of much of the *Clavis Cœlestis*, the writing is diffuse and theologically-orientated; but, as Wright insists repeatedly, it is directed to one central problem, the explanation of the Milky Way. This intriguing phenomenon had not been mentioned in the *Clavis Cœlestis*, but in writing the *Original Theory* he admits that 'This luminous Circle has often engrossed my Thoughts, and of late has taken up all my idle Hours' (p.37). Can we recover something of the train of his thought as he worried his way to a solution of the mystery?

Wright himself provides the key on p.1 of the *Original Theory*, where he tells us that 'The Hypothesis upon which this new Astronomy is founded, and now reduced into a regular System, was the result of my Astronomical Studies full fifteen years ago'. In a footnote he is more specific: 'The first Scheme of this Hypothesis was plann'd in the Year 1734, representing in a Section of the Creation, eighteen Feet long and one broad, several thousand Worlds and Systems, and a great Number of emblematical Figures. . . .' From this two significant facts emerge: first, in 1734 Wright was producing a vast visual aid in the form of a flat, two-dimensional drawing, and this, as we might expect, showed a flat, two-dimensional object, a plane section of the Creation; second, from the talk of emblematical figures, we learn that Wright was by no means engaged on pure science.

Fortunately, although the scheme itself seems not to have survived, a pair of documents related to it have recently come to light and are published for the first time in this volume. With their aid we shall now essay a tentative reconstruction of the development of Wright's thinking about the Milky Way. But as we do this we shall realise that in the general context of Wright's life-work the explanation of the Milky Way is a side-issue, temporarily given prominence in 1750; and that the attempts to give an integrated picture of the moral and physical universe, so often deplored by critics of the *Original Theory*, represent the permanent elements in his cosmological thinking. We shall then be less surprised to find that in later life these attempts led him to a radical reappraisal in the course of which, in a mere sentence or two, he offers an utterly different explanation for the Milky Way.[13]

The first document of the pair was once headed 'The Elements of Existence or a Theory of Eternity', but 'Theory of Eternity' has been altered to 'Theory of the Universe'. It is so brief as to be little more than a descriptive title to the scheme, which it says is

a section of ye Univers . . . comprehending first ye Paradise of imortal spirits in there several Degrees of Glory, surounding the Sacred Throne of Omnipotence. Secondly, the Gulfe of Time or Region of Mortality, in which all sensible beings such as ye planetary bodies are imagind to circumvolve in all maner of direction round the Devine Presence, or ye Eternale Eye of Providence. Thirdly, the shades of Darkness & Dispare supposed to be the Desolate Regions of ye Damn^d^.

On the next page we find a note: 'Wrote in ye year 1734: the author being then 22 years old.'

The second document is a much longer 'explanation' of this *Theory of the Universe* and reads like the text of a lecture discussing the scheme. The importance of the lecture (if such it be) is indicated by the note added at the end:

> This Juvenil Performance was the Produce of the authors Imagination before he had Reap'd any advantages either from Reading or Study, but Prov'd afterwards the foundation of his Theory of ye Univers a much more perfect Work.

But here Wright doth protest too much, for in 1729, when he noted in his Journal 'Conseive may find Ideas of ye Deaty and Creation', his father had recently burned his books because Wright was so madly devoted to the study of astronomy. Before discussing the *Theory of the Universe*, then, let us examine some of the ideas current in books he is likely to have read at this impressionable age.

Among the books on astronomy which enjoyed a wide circulation in English at this time were Christiaan Huygens's *Cosmotheoros* (Hagæ-Comitum, 1698), of which English editions appeared under the title *The Celestial Worlds Discover'd* 1698 and again in 1722; David Gregory's *Astronomiæ Physicæ et Geometricæ Elementa* (Oxford, 1702), published in English as *The Elements of Physical and Geometrical Astronomy* in 1715 and 1726; William Whiston's *Prælectiones Astronomicæ* (Cambridge, 1707), translated as *Astronomical Lectures* and published in 1715 and 1728; John Keill's lectures *Introductio ad Veram Astronomiam* (Oxford, 1718), translated as *An Introduction to the True Astronomy* and published in 1721, 1730 and subsequently; and William Derham's *Astrotheology: or, a Demonstration of the Being and Attributes of God, from a Survey of the Heavens*, published in London in 1715 and many times subsequently. To these we should add Whiston's *Astronomical Principles of Religion* (London, 1717 and 1725), on which Wright drew extensively when writing *Clavis Cœlestis*, and two works carrying all the prestige of Isaac Newton: the English translations of his *Principia* (London, 1729) and the much more manageable *System of the World* (London, 1728).

From these works Wright would learn that the stars are self-luminous like the Sun and probably have their own planets and comets. Most authors (but not Whiston) would go further and say that the stars are similar in size to the Sun, and some set out methods (due to James Gregory and Huygens) of estimating the distances of stars on the assumption that they appear fainter than the Sun only because they are more distant.

Most of the books would tell Wright that the visible stars are only a part of the total creation, and perhaps an altogether insignificant part. But Huygens thinks it possible that 'beyond such a determinate Space [of stars] he [God] has left an infinite Vacuum';[14] and Whiston, in a phrase that could almost have been written by Wright, speaks of 'that unfathomable Abyss of infinite Extramundane Space'.[15] As to the arrangement of the stars, suggestions that they are regularly spaced out are perhaps less significant than the view of Whiston and Derham that the stars appear disordered only because of our position as observers. For Whiston 'it is very rational to conclude, that some regular Order hath Place also amongst the Fixed Stars. There may be a certain orderly and harmonious Disposition of the Fixed Stars amongst themselves, when they are beheld from some other proper Place, altho' that Order appears not when they are seen from this Earth.'[16] To Derham the stars

look to us, who can have no regular prospect of their positions, as if placed without any Order: like as we should judge of an Army of orderly, well disciplined Soldiers, at a distance, which would appear to us in a confused manner, until we came near, and had a regular prospect of them, which we should then find to stand well in rank and file.[17]

But perhaps the most striking passage for our present purpose occurs when Whiston, writing before Halley's discovery that certain stars had a motion of their own,[18] forecasts the danger of destruction of the 'Grand System' of stars:

. . . since withal, the Sun and Fixed Stars do not revolve about one another, or about any common Center of Gravity . . . it follows, that the several Systems, with their several Fixed Stars or Suns, do naturally and constantly, unless a Miraculous Power interposes to hinder it, approach nearer and nearer to the common Center of all their Gravity; and that in a sufficient Number of Years, they will actually meet in the same common Center, to the utter Destruction of the whole Universe.[19]

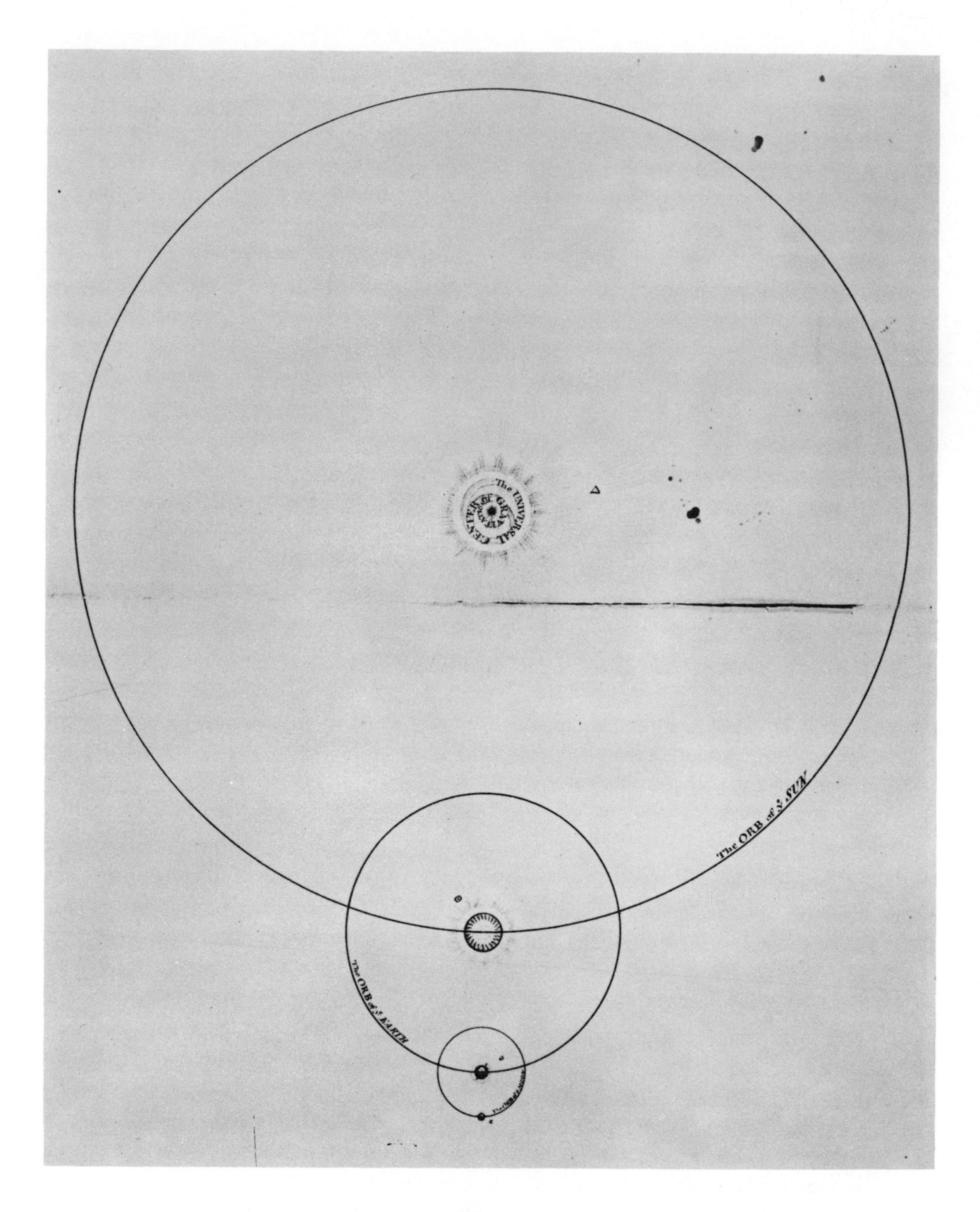

Plate I

Sketch by Wright, presumably for Plate XXI of *An Original Theory*, showing the hierarchy of orbital motions : the Moon in orbit about the Earth, the Earth about the Sun, and the Sun about 'The Universal Center of Gravitation'.

It seems a natural reaction of someone reading this after Halley's discovery, being familiar also with the hierarchical division of Sun, 'primary' planets and 'secondary' planets, and sharing Whiston's concern to integrate the religious and astronomical views of the universe, to suggest that stars of the Grand System *are* in motion about their common centre of gravity, and that this is why the System does not collapse into its centre. Such an orbital motion about the centre is fundamental to Wright's thought in 1734 as in 1750, motion both of the stars in general and of the Sun in particular. Plate 1, a previously unpublished sketch[20] presumably for Plate XXI of the *Original Theory*, clearly indicates the hierarchy of orbital motions and the path of the Sun about 'The Universal Center of Gravitation'. The 1734 description cited above of the 'Gulfe . . . in which all sensible beings such as ye planetary bodies are imagined to circumvolve in all maner of direction round the Devine Presence' is developed at length in the sixth letter of the *Original Theory*. There Wright asks the reader (p.51) to grant that 'all the Stars are, or may be in Motion'; he cites Halley's paper (pp.53-54), and argues that it is by these motions that the stars are prevented 'from rushing all together, by the common universal Law of Gravity' (p.57). But he was more successful in representing the orbital motions of the stars in sketches hitherto unpublished (Plates 2-4) than in the figures of the *Original Theory*.[21]

There was much else in Whiston's *Astronomical Principles of Religion* to appeal to a man of Wright's temperament: talk of 'the noblest or invisible Parts of the Creation', of 'this System of the Universe' being 'God's great House, or Family, or Kingdom', and of the location of Heaven, Hell, and the various spirits.[22] For Whiston, the Devil and his angels, together with wicked men, will one day be transferred from their prison within the Earth to a comet, there to be tormented in the sight of the Blessed who will inhabit our atmosphere. Only slightly less crude is the famous treatise by Tobias Swinden (cited by Derham), *An Enquiry into the Nature and Place of Hell*.[23] According to Swinden, Hell is in the Sun, which is the centre of the 'orbicular' Creation; around the Sun moves the Earth, and around and outside the Earth are the vast numbers of stars, beyond which is the Throne of God.[24]

Wright's own picture concentrates on Hell as a place of darkness rather than of heat, and it lies outside the circle of stars. In this he may well

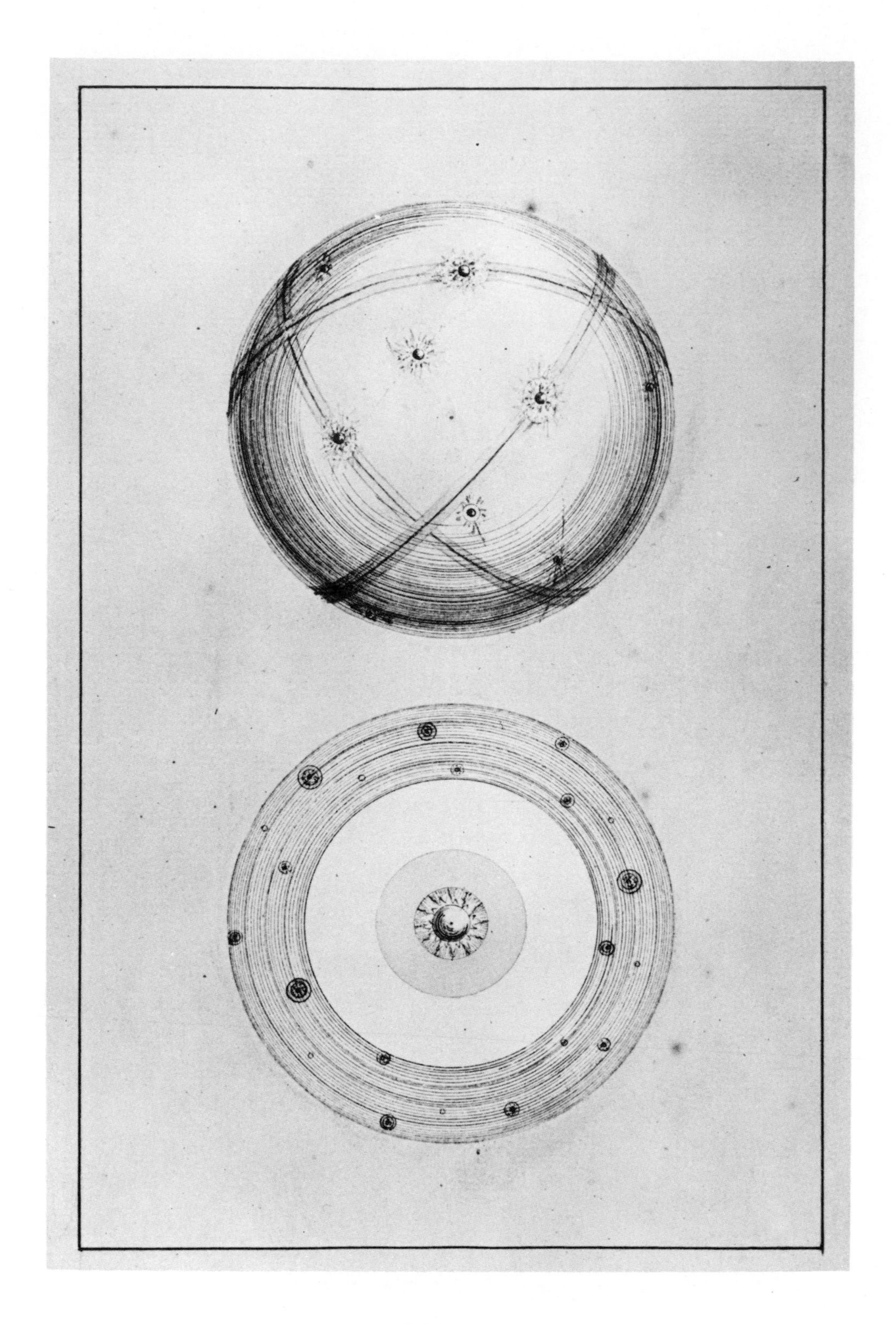

Plate 2
Sketches by Wright, presumably for Plates XXIV and XXV of *An Original Theory*, showing a spherical shell of stars which are orbitting in various directions about the centre.

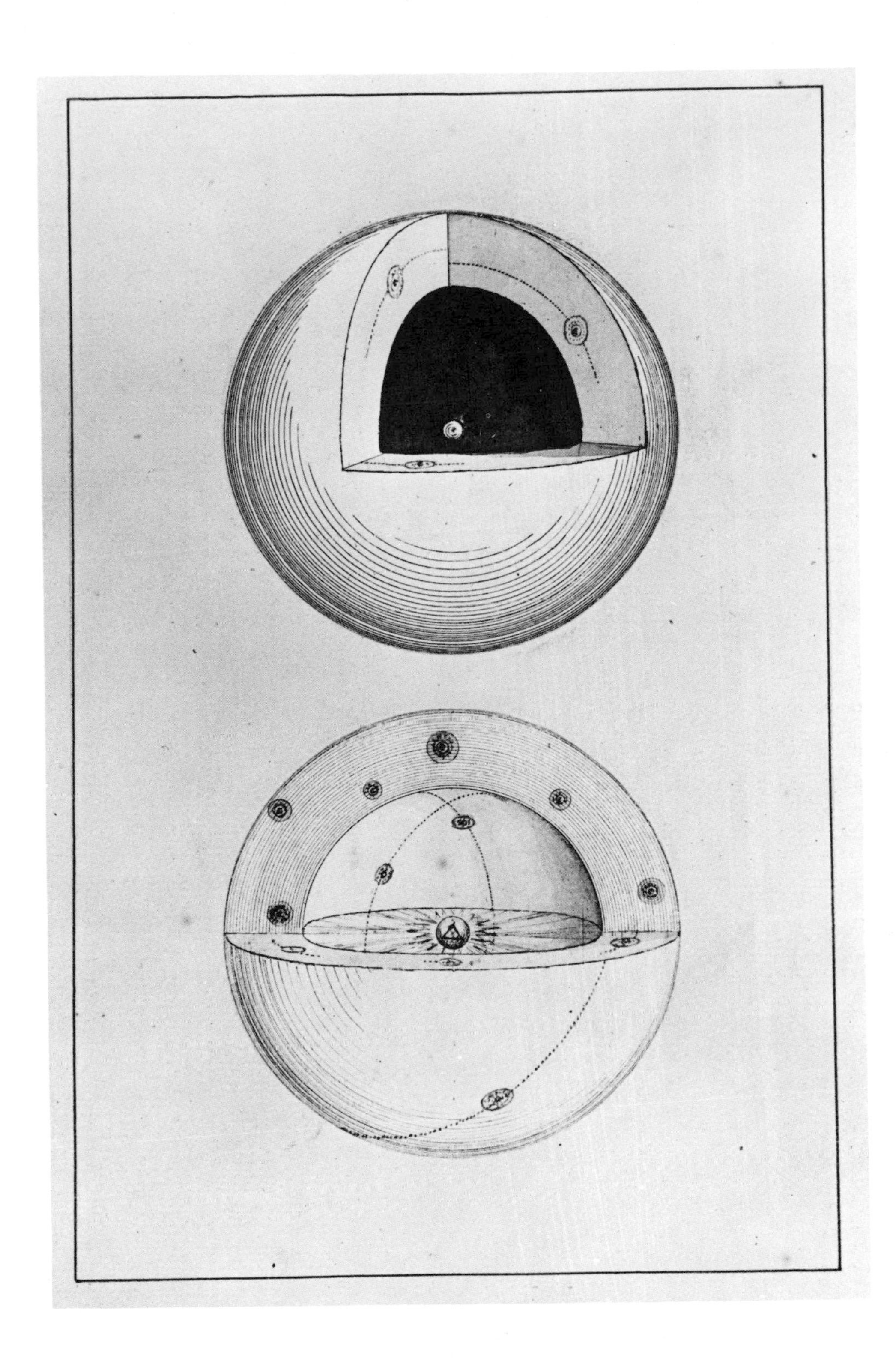

Plate 3

Further sketches by Wright showing a spherical shell of stars which are orbitting in various directions about the centre; the motions of the stars are better represented here than in the published plates of *An Original Theory*.

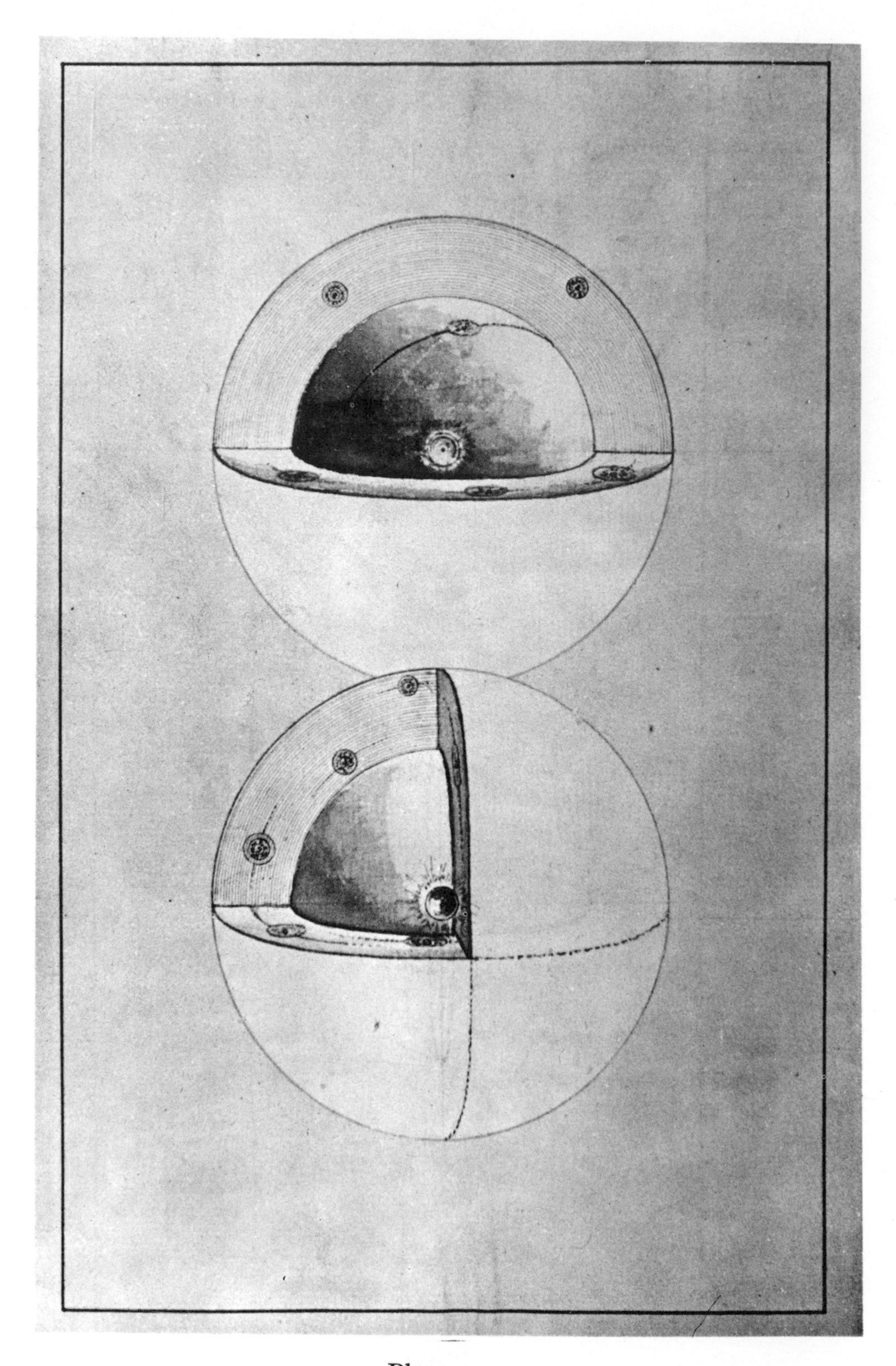

Plate 4
Yet more attempts by Wright to represent the orbital motions of stars.

have been influenced by Robert Fludd,[25] or by the German mystic Jacob Boehme or Behmen whose writings were so popular in early-eighteenth-century London[26] and which contain many elements found in Wright's works: for example, Boehme's *XL Questions Concerning the Soule* tells us of the 'Eye of Eternity', the 'Angelicall World . . . Court of Angels or Princely Thrones in the Liberty of the Divine Majesty', and the 'Abysse of the Darke World';[27] and elsewhere Boehme presents us with the central Eye surrounded by stars which in turn are enclosed by the serpent near to biting its own tail.[28] But Wright's immediate source may well be a more recent work as yet unidentified, in which these ideas are assimilated to the recognition of the Sun as one of the stars.

In this early lecture on his *Theory of the Universe*, we find that Wright is already committed to his life's work in cosmology: the reconciliation of the moral and the physical view of the world. Already he believes *the world has a centre in the moral order*, the Sacred Throne, *which he identifies with the centre in the physical order*, the gravitational centre and source of the laws of nature; and already *the very existence of such a centre is for him a guarantee that both morally and physically the creation is arranged symmetrically around this centre*. We might here interpose the question, Is such a reconciliation worth attempting, in view of the fact that our observational knowledge of the physical order is confined to one little corner of creation? Yes, precisely because of the symmetry about the centre; for, as Wright will argue at length in the second letter of the *Original Theory*, 'from a very small part of orbicular things, we are able to determine the form and direction of the whole' (p. 11).[28a]

Wright was to express these fundamental tenets in a different way in each of his three cosmological writings. In the 1734 *Theory of the Universe* he does it in terms of two spheres, both centred on the Sacred Throne. Inside the smaller sphere is the Region of Philisity or Heaven; outside the larger sphere is the Region of Punishment or Hell, the outer darkness; while in the spherical shell lying between the two spheres is the Region of Probation, and here the solar system, like the other 'miriads of systems in all manner of dispositions', circles about the Sacred Throne.

To add force to his message Wright proceeds to take his audience on a tour of these various regions, after the manner of Keill, who had invited his readers to 'suppose we have the Power of Travelling every where, thro' the Immense Regions of indefinite Space'.[29] For example, some parts of Hell we find are 'faintly enlightened' by the light from our stars, while from other, nearer parts we can see the stars (no doubt by analogy with nebulae) as forming 'a bright and glorious creation'.

Now Wright is very far from being the devout but critical scientist eager, in Whiston's phrase, to compare these two Divine Volumes of revelation and natural knowledge. But to carry conviction he must make some show of relating his overall conception to what we know by observation of our particular corner of the universe, 'demonstrating', as he says in his synopsis, 'ye visible creation to be only a finite view of ye general frame of Nature'. And so, in his scheme (C3 & 4)

> upon all sides the principal stars of the visible creation are exhibited in ther natural order as seen from ye Earth by ye naked eye. Those of ye first magnitude nearest to our own system, and the rest proportionable removed according to their respective phenomena. Beyond these are others more remote crowned by a penumbral shaddow such as we call telescopic stars and again without them more, suposd to be at immense distance & by no means perceptible to ye human eye. At a certain distance from ye Sun equal to a vissual ray of ye smallest visible star is a faint circle of light terminating the utmost extent of ye visible creation, in a finite view from ye Earth . . . .

This explanation of the Milky Way, the 'faint circle of light', is momentous. But Wright's attempt to reconcile observational evidence with his overall theological picture is no more than cursory. The plane section represented in his 'scheme' he defines as 'passing thro' the center of ye [*S*]*edes Beatorum* & coinciding with that of our Earth or ye solar orbit'(B1), and he also says it is 'parallel to ye *Orbis Magnus*' (C3). Is then this versatile plane to contain the Sacred Throne or Sedes Beatorum *and* the Milky Way *and* the ecliptic? No doubt we must make full allowance for artistic license; but this must not blind us, as it has blinded Wright, to one fundamental error: the Milky Way is unique, whereas a faint circle of light should be seen in any *and every* plane drawn through the position of an observer located within 'miriads of systems in all manner of dispositions'. In other words, if the stars were scattered around us in all directions then the *entire* sky would seem ablaze with a faint light.

Whether Wright had realised this by 1742, when his *Clavis Cœlestis* appeared, is doubtful. We would know from internal evidence alone that he had not solved the problem, for he describes the stars as 'promiscuously distributed through the Mundane Space'.[30] But if the Milky Way is the effect of very many distant stars concentrated in certain directions only, the true distribution cannot be promiscuous. By 1750 he had realised this: as he tells us on p. 30 of the *Original Theory*, the stars are 'distributed either promiscuously, or in some regular Order', and his task is to show that the latter alternative is correct.

In the famous middle section of Letter the Seventh, Wright invites us to 'imagine' the stars as scattered in the region bounded by two parallel planes. An observer within this region looking along the layer would see bright near stars, then fainter and more distant stars, and then still fainter and more distant stars, in such numbers that their light would appear to 'form a perfect Zone of Light; this I take to be the real Case, and the true Nature of our *Milky Way*' (p.63). By contrast, the same observer looking *away* from the layer of stars would see only a few bright stars before his gaze reached empty space.

This explanation of the Milky Way as an optical effect due to our immersion in a layer of stars constitutes Wright's chief claim to fame, and the accompanying Plate (XXIII, engraved in error as XXI) is among the most familiar in the history of astronomy. But so far Wright has discussed only a simplified model for the distribution of the *visible* stars, and as he himself immediately says: 'But now to apply this Hypothesis to our present Purpose, and reconcile it to our Ideas of a circular Creation' (p.63). He has still to reconcile this scientific hypothesis that the visible stars form a flattish layer with his theological and moral world picture which by 1750 he had modified and extended but by no means abandoned —a picture in which our star system taken as a whole surrounds its supernatural centre. How was he to do this?

With the unreal clarity that hindsight allows, we see that he could have reasoned somewhat as follows. In 1734 he had drawn a plane section passing through both the supernatural centre and the location of the observer in the solar system. The plane section, on the 1734 world view, could indeed have been any one of the infinitely-many such planes, and yet always it would have contained a milky way. Since the Milky Way is in fact unique and lies in a unique plane passing through the location of the observer, it can actually belong to at most one, and perhaps to none, of these infinitely-many planes.

If it belongs to *one* such plane—that is, if the stars are specially related to a single plane through the supernatural centre—then, according to Wright, our star system *as a whole* is analogous to the rings of Saturn (see Plates XXVIII and XXIX) though the *visible* stars occupy a disk-shaped region in one corner of the ring. On the other hand, if *no* such plane is singled out to contain the Milky Way (and this, with its greater symmetry, clearly appears to be Wright's preferred alternative), then the annulus of stars which lay in the 1734 plane section should have been drawn as too thin to produce a 'faint circle of light' all around the observer. The observer will then see a milky effect in this plane only when he looks tangentially along the annulus. And since the annulus is thin, it follows that the spherical shell of stars of which it is a section is likewise thin. In other words, our star system forms a thin spherical shell surrounding the supernatural centre on all sides; the plane of the Milky Way is the tangent plane to the sphere at the point where the observer is located; and the visible stars are part of a gently-curving spherical layer.

Wright did not find this conception easy to represent, though he devotes four plates to the subject (nos. XXIV–XXVII), and made several other sketches (Plates 2-4). They show that, although Wright will now allow more elements into his universe in the form of multiple or additional systems of stars, on this preferred alternative the 1734 picture persists as essentially the true representation of our own star-system. Nothing would indicate more clearly just how far Wright is from the *modern* conception of our star-system or Galaxy, although it is of course true that his simplified model of the distribution of the *visible* stars has stood the test of time. Strangely, this latter achievement has succeeded almost in obliterating knowledge both of his overall picture of our star-system and of the theological motives and framework to which he was dedicated. As a result not only is Wright often credited with a disk- or lens-like stellar universe, but he has even been described as 'the first astronomer of the "new philosophy" to tackle the problem of the structure of the universe'.

This situation has arisen from a curious set of circumstances. Because of the rarity of the *Original Theory*,[31] Wright's work was known for many years mainly from references in Kant's *Allgemeine Naturgeschichte und*

Theorie des Himmels (1755, although distribution was blighted by the bankruptcy of the publisher).[32] Kant in turn acquired his knowledge of Wright's *Original Theory* from a long review which appeared in 1751 in a Hamburg journal.[33] This account is fair and contains extended quotations, especially of the middle section of the seventh letter, but the two alternative pictures of our complete star system are reduced to a sentence apiece, and the supernatural centre appears to be the centre of creation as a whole and not the centre of our star system. In the inevitable absence of the plates, which show not only that the (local) supernatural centre is in fact in the midst of our stars but also that Wright would have a very great gap between opposite sides of our star system to allow plenty of room for this supernatural centre, Kant not unreasonably thought he was being offered two alternative pictures of our star system, both on a purely natural level. The spherical-shell alternative he felt he could exclude on observational grounds, since he gathered from Maupertuis[34] that nebulae (which Kant believed to be other star systems like our own) appeared elliptiform and not circular. There remained the Saturn's-rings alternative, described in the review by analogy with the solar system, the stars 'all moving in the same way, and not much deviating from the same plane, as the planets in their heliocentric motion do round the solar body'.[35] This analogy greatly appealed to Kant; he was convinced it was sound and that the stars of our system occupy a disk-like region, and he was happy to acknowledge to posterity the debt he imagined he owed to Thomas Wright for this conception.

The next to hit on the same idea was J. H. Lambert, whose *Cosmologische Briefe über die Einrichtung des Weltbaues* was published in Augsburg in 1761. According to a letter he wrote to Kant in 1765 (and there is no reason to doubt its truth), Lambert first outlined his thoughts on the subject in 1749. It was only in 1761, after the composition of the *Briefe*, that he 'was told at Nürnberg that some years previously an Englishman had printed similar thoughts in letters to certain persons, but that he had not had much success, and that the translation of his letters, begun at Nürnberg, had not been completed'.[36]

The concept passed into the mainstream of technical astronomy through the great papers on 'the construction of the heavens' published in the 1780s by William Herschel.[37] Herschel did not, as is sometimes stated, 'prove' the correctness of the view of the Galaxy as disk-shaped, unless by this we mean merely that he confirmed by actual counts the impression we have that more stars lie in the direction of the Milky Way than in other directions. Granted this impression is correct, Herschel's approach was sure to result in a (roughly) disk-shaped Galaxy, since he assumed that the stars are regularly distributed and that his telescope could penetrate to the borders of the Galaxy in every direction. Herschel's triumph was rather to demonstrate the possibilities of stellar statistics and to promote the whole question of the Milky Way beyond the limits of the merely speculative; but it was open to later astronomers to question the validity of his assumptions (as indeed Herschel himself did in later life), and to suggest instead that the Milky Way is what it looks to be: a ring of small stars surrounding a collection of larger stars which includes our Sun.

But did Herschel derive his interest in the problem from Wright, or Kant, or Lambert, or did he hit upon it independently? Probably we shall never know: as a boy in Hanover Herschel had been introduced to astronomy, but he came to England in 1757 as a refugee after Hanover was occupied by the French, and he can hardly have then known of the copies of Kant's work appropriated from the bankrupt publisher; on the other hand he may well have learned of Wright's *Original Theory* in the early 1760s when he travelled extensively in the north-east of England as a music teacher. He makes no mention of Wright, and his copy of the *Original Theory* is little-marked except for a note which could have been written any time after 1781.[38] There are, however, a number of curious ideas common to both the *Original Theory* and Herschel's published papers, which are sufficiently unusual to constitute some evidence that Herschel may have read the *Original Theory* early in his career. For example, Wright's proposal (pp. 55–56) that changes in stellar magnitudes may be caused by movements of stars is echoed in a 1783 draft where Herschel speaks of 'changes in the apparent magnitudes and reciprocal distances of the stars';[39] and Wright's suggestion (p. 45), quickly withdrawn, that stellar distances are directly proportional to their apparent magnitudes as classified by astronomers was an hypothesis used by Herschel for most of his career. Yet both were sufficiently implausible to arouse most strenuous resistance from Nevil Maskelyne when incorporated in papers submitted to the Royal Society.[40]

In the nineteenth century Wright continued to be known mainly through the references in Kant, although an undistinguished edition of the *Original Theory* was published without the plates in 1837 in the U.S.A.[41] The relevant plates were again missing from the 'Account of the Speculations of Thomas Wright of Durham' given by Augustus De Morgan in 1848, although (unlike the Hamburg reviewer) he does quote the later parts of Letter the Seventh in full.[42] The text references to Saturn's rings as the analogy for one model of our star system should, one might think, have been clear enough even in the absence of the plates. But De Morgan too makes the fatal transition from the visible stars to our complete star system. Wright, he correctly says, 'gives the explanation of the phænomena of the milky way, as now generally received',[43] but later varies this to 'He gave the theory of the milky way which is now considered as established, contended for what is now called the central sun . . .'.[44] For Kant and his successors this was, as we have seen, an easy transition, encouraged by the partial account of Wright's views that alone was available to them; the failure of De Morgan, and others with access to the *Original Theory*, to recognise the misapprehensions of earlier authors has done justice neither to Wright's own thought nor to Kant's distinctive contribution.[45]

If we now turn our attention from Wright's treatment of the Milky Way and survey the *Original Theory* as a whole, we find that in its overall arrangement, the book is irritatingly diffuse: the blame for much of the misunderstanding with which Wright's work has been surrounded must rest with him. Yet to deplore his religious motives, and to regret the space he devotes to poetry and to history (in which he was greatly indebted to Sherburn's rambling *Sphere of Manilius*)[40] is anachronistic, as the first letter illustrates: a reverential and poetic approach to Nature was common in that more spacious age, and severely technical textbooks of astronomy were out-numbered by works inspired by a more humane outlook.

In Letter the Second, Wright justifies his future use of analogy, as for example when he will assume the stars have their own planetary systems; and he shows how, when we study objects with circular or spherical symmetry, we can reason from the part to the whole. He then with slight excuse interpolates a summary of primitive conceptions of the heavenly spheres.

The third letter gives details of the solar system, and includes the suggestion (p.21) that the orbits of all comets may be equal in area. Here he prints calculations entered in manuscript in his own copy of *Clavis Cœlestis*, now in Durham University Library. Wright's suggestion is quoted at length by the Hamburg reviewer and described as 'a very happy remark', but De Morgan, writing a century later and looking for elements of Wright's *Original Theory* still scientifically acceptable, takes the very opposite view: 'This is not a happy conjecture.'[47]

Letter the Fourth is notable, as we have seen, for containing the germ of Wright's insight into the structure of the Milky Way: that the stars are distributed, not at random, but 'in a regular Order' (p.30). Wright's reference in the same letter to 'the *known* Planets' (p.31, italics ours) has been praised as an anticipation of the discovery of Uranus. But he is merely repeating a phrase used by Whiston in 1717 in the *Astronomical Principles* and by himself in 1742 in *Clavis Cœlestis*, and the possibility of planets beyond Saturn is elaborated at length by William Wall in 1727 in the second edition of Swinden's *An Enquiry into the Nature and Place of Hell*.[48]

By contrast, a recent critic has with strange perversity compared Wright unfavourably with Kant for advocating, here and elsewhere, the 'queer idea' that the gravity of each star system was limited. We must not forget that at this time there was no proof that gravity did extend beyond the solar system: not until 1767 did John Michell show that multiple stars occur far more often than could be expected if no physical causes were at work;[49] not until 1803 did Herschel give observational evidence that pairs of stars were in orbit around each other under attractive forces, whatever they might be;[50] and it was decades more before these forces were identified as gravitational. But in fact Wright's critic has fastened too easily on a single sentence. Newton (cf. *Original Theory*, p.6) and many subsequent authors had contrasted the effective operation of the Sun's gravitational attraction on planets and comets with the negligible consequences of the attractive forces of other stars. Wright, in spite of his mysticism a very practical man, concentrates on the '*sensible* Sphere of the Sun's Attraction' (p. 31), on 'a Space sufficient to divide or seperate the *sensible* activity of neighbouring Systems' (p. 70, all italics ours), and by so concentrating lapses into occasional ambiguity. His mode of thought is clear from a section of the *Clavis Cœlestis*, where he says of two bodies: 'If the Body C had no natural tendency to the Body A, or more properly, if the Body A were removed to an infinite Distance from that of C, so as to render its Virtue of no Effect at C . . .'.[51] In the *Original Theory*, as in the 1734 lecture, Wright has the stars in projectile motion about a centre and it is this that prevents them 'from rushing all together, by the common Universal Law of Gravity' (p. 57).[52]

Letter the Fifth gives Wright the excuse to include some of his most delicate illustrations, based in part on his own modest attainments as an observer. His ideas on the distances separating stars, and on the number of stars visible in the Milky Way, are conservative enough, but, even so, combined they reveal a vastness in the visible creation the excitement of which we can still recapture. Significant for the transformations that overcame Wright's ideas in later life is his account (p. 43) of 'the astonishing Phenomenon of several new Stars', a topic which was to play a crucial role in his *Second Thoughts*. The letter closes with a hint (p. 46) of 'a general Motion of [the stars] round a common Center', and with an attempt to whet our appetite for the future happiness awaiting us among the stars and in 'the final Coalition' in the Abode of the Blessed.

In the sixth letter, Wright argues that the stars are individually in motion, and that it is these projectile forces that prevent the system from collapsing. This analogy with the planets (and with the 'lesser planets' he believed formed Saturn's rings) he continues in Letter the Seventh; and just as the planets appear erratic because of our eccentric position,

> we may readily imagine, that nothing but a like excentric Position of the Stars could any way produce such an apparently promiscuous Difference in such otherwise regular Bodies. And that in like manner, as the Planets would, if viewed from the Sun, there may be one Place in the Universe to which their Order and primary Motions must appear most regular and most beautiful. Such a point, I may presume, is not unnatural to be supposed, altho' hitherto we have not been able to produce any absolute Proof of it. See *Plate* XXV [which shows a section of our spherical creation, an annulus of stars 'with the Eye of Providence seated in the Center']. This is the great Order of Nature, which I shall now endeavour to prove and thereby solve the Phænomena of the *Via Lactea*.[53]

This passage alone would be sufficient to demonstrate that Wright plans to integrate his explanation of the visible Milky Way (or, rather, his first simplified model for such an explanation), *which immediately follows*, into his overall conception of the stars as surrounding the supernatural centre; and that it was therefore utterly impossible for him to propose a grindstone theory of our star system.

We have already discussed how this integration was achieved in the remainder of the letter, and why Kant and subsequent writers failed to realise what Wright was about. The final two letters might have made Wright's position clear, had his readers not been discouraged by their philosophical titles. Letter the Eighth, for example, contains an estimate (p. 73) of the scale of our star system: the shell (or ring?) has a thickness of only some eight million million miles, but an inner diameter of anything from twenty-six to ninety-six million million miles; and a star may take over a million years to orbit around the supernatural centre (p. 75). Letter the Ninth takes it as granted 'that the Creation may be circular or orbicular' (p. 78) and 'that all the Stars may move round one common Center' (p. 79), and enquires into the precise nature of this supernatural centre. In this section, more than anywhere else, Wright's unification of the natural and the supernatural, of efficient causes and final causes, is clearly expressed:

> To this common Center of Gravitation, which may be supposed to attract all Vertues, and repel all Vice, all Beings as to Perfection may tend; and from hence all Bodies first derive their Spring of Action, and are directed in their various Motions.
>
> Thus in the *Focus*, or Center of Creation, I would willingly introduce a primitive Fountain, perpetually overflowing with divine Grace, from whence all the Laws of Nature have their Origin . . . .[54]

In the final paragraphs, Wright supposes that 'the endless Immensity is an unlimited Plenum of Creations not unlike the known Universe' (p. 83) and that the nebulae may be just these. He has several times (pp. 41, 42, 65) suggested that not only the Milky Way but other 'luminous spaces' in the heavens are the effect of large numbers of stars; but on p. 65 he regarded them not as independent systems but as possible components of a compound system to which our stars belong. Wright had evidently not thought out the implications of this proposal. For example, if our Sun is a member of a spherical shell of stars which we suppose to form the innermost component of our compound system, then the sky to one side of the Milky Way (in the direction of the supernatural centre) ought to be empty of nebulae; if it is the outermost component then the nebulae visible to one side of the Milky Way would be other components, of which all except perhaps the innermost would necessarily be in the form of rings rather than shells of stars, while nebulae visible on the other side would be independent star systems.[55]

But granted there are other 'creations' like our own star system, what, one wonders, prevented Wright from carrying the hierarchy of moon-planet-star-creation one step further and, like Kant and Lambert, imagining these creations to be themselves components of a still higher system? Were the creations wholly in the natural order, this step would be an easy one. As it is, the question must be answered by counter-questions: What meaning could one possibly give to an hierarchy in the supernatural order? How can one supernatural centre be satellite to another? No, for Wright the way ahead is blocked by the very multiplicity of Abodes of the Blessed, and the next step in the elaboration of systems of systems had to be left to thinkers for whom the natural and the supernatural are less closely aligned.

NOTE ON 'A THEORY OF THE UNIVERSE'

Many of Wright's effects were purchased by Mr George Allan of Darlington, and when books and manuscripts from Allan's library were auctioned by Sotheby's on 18 December 1844, they included numerous papers of Wright, among them many if not all of the manuscripts now in Newcastle Central Library. These were purchased by Newcastle in 1898 from a Mr H. Gray for £3. *A Theory of the Universe* is cleanly written and without pagination, but every fourth page is headed with a new letter of the alphabet; here the page headed 'A' is indicated by 'A1' and the pages following by 'A2', 'A3' and 'A4', and so on. Interpolations by Wright, either in revision or at a later date, are placed within angle brackets (< ,>), and editorial emendations within square brackets ([,]). Wright's capitals and punctuation, which are often ambiguous and seldom significant, have been freely altered. Grateful thanks are due to the City Librarian, Newcastle, for kind permission to publish this manuscript.

NOTES

1. Much of our knowledge of the first half of Wright's life comes from his own journal (British Museum ADD. MSS. 15627), published by Edward Hughes in 'The early journal of Thomas Wright of Durham'(*Annals of Science* **7** (1951), pp. 1–24). The quotations which follow are taken from this journal. Much bibliographical and other information about Wright was collected by the late F. A. Paneth and is conveniently assembled in *Chemistry and Beyond*, a selection of Paneth's writings edited by Herbert Dingle and G. R. Martin with assistance of Eva Paneth (New York, 1964). Paneth's account of Wright's theories, however, must be used with great caution.

2. See for example the list of William Whiston's publications at the end of his *Astronomical Principles of Religion, Natural and Reveal'd* (2nd edn, London, 1725).

3. Copies of the text are common, but the complete scheme is excessively rare. A facsimile of the text and scheme, with introduction by the present writer, was recently published (London, 1967).

4. For details of Wright's published (and some unpublished) writings, see *Chemistry and Beyond*, pp. 114–19. A list of Wright's manuscripts is being prepared by M. A. Hoskin and D. M. Knight.

5. *Op. cit.*, p. 2.

6. *Chemistry and Beyond*, p. 114.

7. According to the title page.

8. Copies survive in Durham University Library. I am grateful to the Librarian and to Dr A. I. Doyle for allowing me access to their rich collection of Wright material.

9. *Clavis Cœlestis* (London, 1742), p. ix.

10. Ibid., p. 16.

11. Ibid., p. 75.

12. In the library of the Royal Astronomical Society. The Table of Contents differs somewhat from the published version and corresponds to an earlier organisation of the material. Letter IV as published appears to be a late addition, and Letter V to be an amalgamation of the original Letters IV and V; for in the prospectus Letter IV is 'Of the Order, Distance, and Multiplicity of the Stars, within our finite View', and Letter V is 'Of the *Via Lactea*, &c. proving the known sidereal Creation to be finite'. Most of the other changes are trivial, but in the published version 'amongst the Stars' has been added to the title of Letter VI and 'proving the sidereal Creation to be finite' to that of Letter VII.

13. Attention was first drawn to these documents and to their theological nature by Dr D. M. Knight of Durham University in 1965; see *Actes du XIe Congrès International d'Histoire des Sciences* **3** (Warsaw, 1968), pp. 37–40.

In later life Wright returned once more to the problem of integrating the natural and the supernatural, in *Second or Singular Thoughts upon the Theory of the Universe*, a sequel to the *Original Theory* in the form of three further letters and other material. The manuscript of this work came to light only in 1966 when Mssrs Dawsons of Pall Mall invited the present writer to help sort the chaotic mass of Wright material they had purchased at auction. In these *Second Thoughts*, Wright abandons much of his earlier doctrine, and now teaches that the stars are celestial volcanos set in a heavenly shell which surrounds the Sun! This manuscript is now in the possession of Durham University; an edition by the present writer was recently published (London, 1968).

14. C. Huygens, *The Celestial Worlds Discover'd* (2nd English edn, London, 1722), p. 157.

15. Whiston, *Astronomical Principles*, p. 122.

16. Whiston, *Astronomical Lectures* (2nd English edn, London, 1728), p. 42.

17. W. Derham, *Astrotheology* (London, 1715), p. 55. Cf. Wright, *Original Theory*, p. 50: 'we may reasonably expect, that the *Via Lactea* . . . will prove at last the Whole to be together a vast and glorious regular Production of Beings . . . and that all its Irregularities are only such as naturally arise from our excentric View'; and ibid., p. 62: '. . . in like manner, as the Planets would, if viewed from the Sun, there may be one Place in the Universe to which their [the stars'] Order and primary Motions must appear regular and most beautiful'.

18. E. Halley, 'Considerations on the Change of the Latitudes of some of the principal fixt Stars' (*Phil. Trans.* **30** (1717–19), pp. 736–8).

19. Whiston, *Astronomical Principles*, pp. 88–9; cf. Whiston, 'A discourse . . . of the Mosaic history of the Creation', p. 37, in his *A New Theory of the Earth* (London, 1696); the question had of course been discussed by Richard Bentley and Isaac Newton in 1692 and 1693 (*The Correspondence of Isaac Newton*, edited by H. W. Turnbull, **3** (Cambridge, 1961), letters 398, 399, 403, 405 and 406), and the present writer is preparing an edition of related Newton manuscripts.

20. I am grateful to the Curators of Durham University Library for permission to reproduce Plates 1–3.

21. I am grateful to the Council of the Royal Astronomical Society for permission to reproduce Plate 4.

22. Whiston, *Astronomical Principles*, pp. 25, 131–2, 154 seqq.

23. T. Swindon, *An Enquiry into the Nature and Place of Hell* (London, 1714). For earlier ideas on the subject, and especially those of Milton whom Wright so often quotes, see C. A. Patrides, *Milton and the Christian Tradition* (London, 1966), pp. 280 seqq.

24. Swinden, *An Enquiry*, pp. 225 seqq.

25. Cf. Robert Fludd, *Utriusque Cosmi Maioris scilicet Minoris Metaphysica* (Oppenheim, 1617), pp. 131, 141. I first owed this suggestion to Professor Walter Pagel; as Fludd was little read at this period, the suggestion might have seemed implausible, but the subsequent discovery of Wright's *Second Thoughts* with its reference (Letter II, pp. 15–16; 1968 edition, p. 51) to Fludd's *Tomus Secondus de Microcosmi Historia* (Oppenheim, 1619) shows that at least in later life he knew something of Fludd's work.

26. Cf. Nils Thune, *The Behmenists and the Philadelphians* (Uppsala, 1948). Richard Roach, leader of the Philadelphians, was Rector of St Augustine's, Hackney, until his death in 1730, the year in which Wright worked in London for makers of mathematical instruments. For contemporary ideas of Hell see D. P. Walker, *The Decline of Hell* (London, 1964).

27. Jacob Behmen, *XL Questions Concerning the Soule* (London, 1647).

28. Boehme, *Alle Theosophische Schrifften* (Leipzig?, 1730), *Christosophia* plate 7.

28a. A survey of the development of Wright's cosmological thought as expressed in these three writings is given in Michael Hoskin, 'The Cosmology of Thomas Wright of Durham', *Journal for the History of Astronomy* **1** (1970), pp. 44–52.

29. J. Keill, *Introduction* (1st English edn, London, 1721), p. 17.

30. *Clavis Cœlestis*, p. 75.

31. A copy of which was recently on sale at $1500.

32. The story is related by W. Hastie in the introduction to his translation of the work and other related pieces, *Kant's Cosmogony* (Glasgow, 1900); a reprint of this volume with introduction by Dr G. J. Whitrow is in preparation, and a partial reprint with an introduction by Professor Milton K. Munitz has recently appeared (Ann Arbor, Mich., 1969).

33. Freye Urtheile, Achtes Jahr (Hamburg, 1751), translated by Hastie, *op. cit.*, Appendix B.

34. P. L. M. de Maupertuis, *Discours sur les différentes figures des astres* (Paris, 1742); Hastie, *op. cit.*, pp. 32–33. For a critical assessment of the evidence, see Kenneth Glyn Jones, 'The observational basis for Kant's *Cosmogony*' (*Journal for the History of Astronomy*, **2** (1971), in press).

35. Hastie, *Kant's Cosmogony*, p. 190.

36. Ibid., p. lxx.

37. Reprinted in M. A. Hoskin, *William Herschel and the Construction of the Heavens* (London, 1963).

38. Ibid., pp. 115–16.

39. Ibid., p. 54.

40. Ibid., pp. 32, 42. But Wright does in fact later assume that a star of the ninth magnitude is nine times further than one of the first magnitude (*Original Theory*, p. 73).

41. *The Universe and the Stars*, being an original theory on the visible creation, founded on the laws of nature, by Thomas Wright, with notes by C. S. Rafinesque (Philadelphia, 1837). In the Dedication to the American Public, Wright is seen as the reviver of the ancient learning of Egypt, and the republication of his book was the first activity of the American Institute of Learning, newly founded 'for the purpose of restoring, spreading and increasing knowledge of all kinds and on all topics. . . . To the members of it will be due the foundation of an Eleutherium of Knowledge or free school of useful knowledge, an extensive mutual library and museum, and the gradual publication of valuable works on all branches of human knowledge' (Preface, p. 5). Rafinesque, who describes himself as Professor of historical and natural sciences, compares Wright to Plato, Copernicus, Newton and Herschel (p. 7); he has some inkling of Wright's conception of our star system, but is dazzled by the brilliance of Wright's religious insight. He apologises for the absence of the plates (p. 8), but promises their separate publication in the future.

42. A. de Morgan, 'An Account of the Speculations of Thomas Wright of Durham' (*Philosophical Magazine* (3) **32** (1848), pp. 241–52), reprinted in Hastie, *Kant's Cosmogony*, Appendix C.

43. *Kant's Cosmogony*, p. 197.

44. Ibid., p. 203.

45. To this we must record one honourable exception, Vera Gushee, whose article 'Thomas Wright of Durham, Astronomer', appeared posthumously in *Isis* in 1941 (**33**, pp. 197–218). The portrait she reproduces is probably of another Thomas Wright, and many other criticisms can be made of her work, but she does realise that Wright taught no disk-theory of the Galaxy. Yet Paneth (*op. cit.*, p. 118) says 'This article fails to understand Wright's *Original Theory*'!

46. Edward Sherburne, *The Sphere of Marcus Manilius* (London, 1675).

47. *Kant's Cosmogony*, pp. 182, 194.

48. Swinden, *An Enquiry* (2nd edn, London, 1727), p. 355; Whiston, *Astronomical Principles*, p. 19; Wright, *Clavis Cœlestis*, pp. 16, 17, 33. Wright discusses the matter further in *Second Thoughts*, Letter I, p. 31 and Letter II, p. 14 (1968 edition, pp. 42, 50).

49. J. Michell, 'An Inquiry into the probable Parallax, and Magnitude of the fixed Stars, from the Quantity of Light which they afford us, and the particular Circumstances of their Situation' (*Phil. Trans.* **57** (1767), pp. 234–64).

50. W. Herschel, 'Account of the changes that have happened during the

last Twenty-five Years, in the relative Situation of Double Stars' (*Phil. Trans.* **93** (1803), pp. 339–82).

51. P. 41.

52. This is brought out especially by Plate I supplied for this edition.

53. *Original Theory*, p. 62.

54. Ibid., p. 79.

55. A hint of these complications is to be found in the negligible criticisms made in an open letter to Wright and signed 'S. L.', *The Gentleman's Magazine* **21** (1751), pp. 315–17.

THE ELEMENTS OF EXISTENCE
OR
A THEORY OF THE UNIVERSE

[Wright MSS Vol. VII, Central Library, Newcastle upon Tyne]

An hypothesis to solve, nearest to the level of ye human understanding, the Being of a God and ye origin of Nature grounded upon visible effects and final causes. Demonstrating not only ye visible creation to <be only> a finite view <of ye general frame of Nature>, but also the general disposition of the whole <is evenced from it.> Together [2] with ye several mansions of hapiness & misary in a future state | and to be rationaly expected after death. All which is represented in a section of ye Univers twelve feet radius, extending from the Imperial Seat or *Sedes Beatorum* to ye verge of chaos bordering upon ye infinite abiss. Comprehending first ye Paradise of imortal spirits in there several Degrees of [3] Glory sourounding the Sacred Throne of Omnipotence. | Secondly, the Gulfe of Time or Region of Mortality, in which all sensible beings such as ye planetary bodies are imagind to circumvolve in all maner of direction round the Devine Precence, or ye Eternale Eye of Providence. Thirdly the shades of Darkness & Dispare supposd to be the Desolate Regions of ye Damnd. |

[4] <Wrote in ye year 1734: the author being then 22 years old.> |

[1] THE EXPLANATION

At the top is a folded snake; the emblem of Eternity, supported by four boys representing the perpetual youth or mysarable [?] state of ye four elements, wch are supos'd to be the active powers of creation and a little below these upon a double label is this general title:

THE THEORY OF EXISTENCE.

On each side of these by two & two are placd four of ye seven arch angels, namely

**Raphael* **Ithuriel* **Iophel* and **Uriel.*

These are represented as ye active agents of God in the image of his Devine Being, and personal both in majesty & power as many corporial Regents. | [2] A little below these on one side, is ye following motto.

The Eyes of ye Lord are in every place,
Beholding ye eval and the good

Prov: XV, 3.

and on ye other side opposite to it; this

Be wise now therefore O ye Kings,
Be learned ye that are Judges of ye Earth.

Psalms: II, 10.

In ye middle betwixt ye above labels is the Seraph *Abdiel* supporting <a Book of> the Holy Scriptures the leaves of which are illuminated by two Suns representing the two great lights of natural knowledge <as flowing> from reveal'd relegion; the Law and the Gospel, in this is written

!*That I may publish with ye voice of thanksgiving*
and tell of all his wondrous works. |

[3] Immediately below this is an historical and emblematical view of the primitive creation as related by *Moses* with ye expulsion of the evall geneouses or diabolical spirits from ye *Emprimum* or primogenial Seat of Good. Here are severall emblematicl figures to express the sacred and devine Vertues of that Eternall Trinity in ye union of which we comprehend the Deaty. The *Simbol* of this Holy Trinity, chiefly consists in a displaceation of ye †Tetrahedron, by which is ment the mistical union of

* Those impressions are all from that [edge?] 4 inches.

† The property of every equilateral triangle is such that ye nearest approach, contact, or union of any three, adequate to each other make one more of the same kind & similar to ye whole. [4] [The] power & propriaty of the figure | is evedent from the expression of one in three & three in one.

those necessary and self existing Principals which together we call the *Elohime*: over this is represented

The Eye of Providence. |

[4] How this will agree with the Trinity of ye Church which you explain in a very different maner, I shall not make it my business here to determine but leave the true senses of it, to be expounded by abler heads.
I have indede indevoured to define it thus

one omnicient; one omnipotent, and one omnipresent Being, unighting as by necessity *constitude one God* <*whose state is*> *infinite & eternall.*

For to have cognisense without power & power without dominion, implyes an universall negative. & again, dominion without power and power without knowledge is still more absurd; and consequent of much more fatal errors, in the idea of an active ignorance. But in the union of ye three we comprehend the perfection of all beings, namely the *Living God.*|

[A1] Now as the antient Egyptians chose to express by characters; significant words, and simbols, the misteries of their relegion; and particularly such conceptions <of ye Deaty> as all the force of language could not <otherwise> express, I have so far followd their examples as to fix the following *Hyroglyphic* in place of ye Devine presence full in the middle of ye principal scheme.

Omnicience.
Omnipresence.
The
Elohime
or Jehovah.
Omnipotence.

What I mean by this is the fountain of all life, joy and love, from whence we derive our being, and all the blessings we enjoy, thrô our existence. & by ye same infinite Wisdom and all mighty Power every where existing, I would willingly understand this part of *St Athenatious Creed*: God of God, Light of Light, very God of very God, before all worlds, incomprehensible. |

[A2] Near to the Emblematic Trigon, upon the right, is a figure of Piety or Devotion, representing Relegion, and on the left the Geneous of Natural

Philosophy, as presidinge over knowledge, and perticular ye science of astronomy, the former attentively listening to ye latter who is supposd to be explaining the disposition & laws of the New Creation. Below these at some distance is shown the revolt or expulsion of Satan & his angels, who are all together represented falling from the Imperiale Heaven of Heavens, Satan alone preserving an erect *atitude*, as resisting the Devine wrath, with this false motto upon his sheald.

——* *There is no God.*

Amongst the falling angels, I have introduced in a prostrate posture of dread & horor Original Sinn, and over her head as descending the Chirub Zophil, one of the host of Heaven, bearing a masy chain to keep those wicked spirits in eternal bondage. Other angels on each side are sound[ing] thro clarion trumpets the everlasting fame of the victorious triumph. |
[A3] Under the falling angels is this motto:

Hell and Distruction is before ye Lord.

Above these diabolical spirits in ye radiating sillia proceeding from the Deaty is a figure of *Urainia* the heavenly muse, seated in a shining cloud and suported by representatives of Trouth & Justice each bearing the propper emblems of Sacred Wisdom and Power. Here are also respectively display'd the mistic Bokes of Fate, containing the decrees and ye perpetual laws of ye Eternal Will.

The Mottos are

Alpha et omeg.

and

He standeth in the Congre[g]ation of ye Mighty
And judgeth amongst the Gods.

Psalms LXXXII 1st. |

[A4] Immediately below this historical view, commences the grand section or rather sector of ye creation, it being a partial radii, passing thrô all the Regions of Philisity, of Probation & Punishment, from the very verge of

* The fool hath said in his heart, there is no God.

the orbicular creation to the Relms of perfect Bliss. And here that part of it peculiar to ye more imediate Presence of God i.e. the visible Deaty, is fully display'd and suported by 6 Gardian Angels, at the top is a figure of Fame sounding the praises of ye Eternale Godhead upon his act of creation.

This end of the scheme comprehending the center, the other as being the extremity of the orb, and supposed to be the most remote from the focus of power, is void of all created beings & is suppor[t]ed by evale geneouses. Here as upon the verge of Chaos surrounded by discord and [B1] confusion is an emblem of extreme disgrace. | This most horrid spectacle is expressd by a human figure upon whose head the Devine vengence is represented falling & breaking like thunder & lightning, and on each side are two demons supposd to be tormentors.

Thus I have given you a general discription of the signifier or section, as the whole is represented & supported 16 feet long, and shall now proceed to a more particul[ar] one.

First, you are to imagin that this particul[ar] view of the Universe, is a section of the visible creation, formed upon a segment of ye infinite plain passing thrô the center of ye *Cedes Beatorum* & coinciding with that of [B2] our Earth or ye solar orbit. | The lucid body, or radient point of this scheme is supposed to be one of those Eternal Mansions appropriated by ye Devine Being for his first born, or Eldest Sons, begotten of his Sacred Will before all Worlds: Deo of Deum. In these imediate or visible Residences of the Deaty from whence he becomes sensible to his cretures and omnipresent thrô all infinity, Might, Majesty & Dominion is made manifest by an en[d]less Sphere of Glory. The center of which being imagined the *Primogenit[al] Fountain* of all perfection from whence the ever active powers of Creation & generation receiv'd there secret Laws I suppose to be the true *Sanctum Sanctorum* of that inatiquate Being of Beings which ye silent soul of every mortal dreads, which Providence makes manifest in ex[is]tence, but what ye human mind is [un]able to [B3] conceive. | Here the Eternall Spirit <comprehensive of all good> was pleasd to promulgate his powerful will by promising of one Sacred all active & inliving Word, productive of all temporal beings, i.e. the generale *Fiat*.

In place of Whome you have this combination

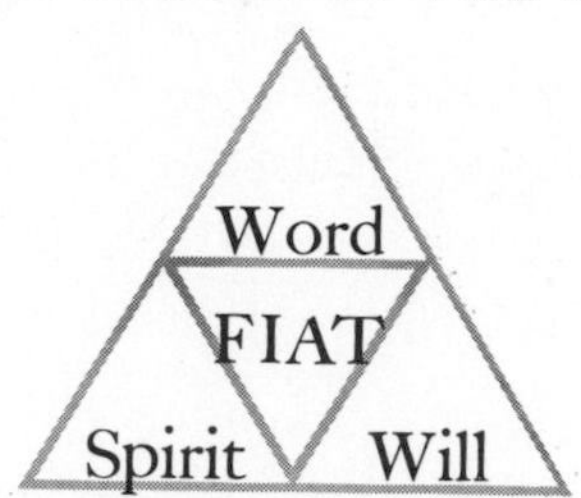

Thus in conformity with y^e Trinity of Etern[ity], the Spirit or Fountain of infinite Light concerning such an effect of thought as from the result of reason, namely the will, this determation of infinite Wisdom imediatly proceeds to generate the primary Cause and lastly by the Word comprehending allmighty Power, the joyous act of creation was made manifest in an en[d]less chain of varigated [?] beings. |

[B4] Hence by a joint adherence of the Spirit to the Will and ye Will to ye Word all firmly unighting in ye act or de[e]d, ye universal laws for a perpetual change are thus promulgated from their Heavenly Origin.

The general section you are to understand is principally devided into three parts which for distinction let me call the internal, ye external & ye middle regions, and in order that we may proceed in ye most regular and natural order, 'twill be necessary to begin our explanation at ye place of our present habitation (viz) the Earth. |

[C1] The system of the Sun round whome ye Earth performs her annual motion is situated near ye center of ye middle region; this for better distinction I would call ye Sphere, or Gulf of Mortality which within its orbicuity comprehends all present beings in their natural state of probation. This stary firmament or co̅mon frame of terrestrial worlds I supose to be bounded by the two extremes of Nature, namely ye desolate Regions of Darkness and ye joy full Relms of Light. The former or external region is supos'd to limit ye human state or creation on ye side of Chaos, this circumscribes the whole & borders upon the *old Eternal Night.* ✓ The latter or internal region is by Creation circumscribed as the common center of gravitation to all the radiant bodies round it, and this I can [C2] easily conceive to be ye true imperial seat | or *Sedes Beatorum*, common to all superior spirits in the most exalted state of happiness.

Betwixt these two ye Gulph of Human Life is fix'd and suppos'd to be so situated betwixt ye Good & Eval Agents, that neither vice nor vertue can originally produce predomenant passions, but yet externally attended with every faculty of sense that may possibly corrupt the primitive inclinations, or add fresh lustre to ye principals of virtue, by exciting the soul or new moduling the mind, according to ye dictates of Devine Reason.

<This tho a state of perfect inocense & freedom is> equally open on all sides both to ye influence of Heaven and ye insults of Hell, obnoctious and exposed for a trial of vertue, originaly designd for human good and finaly the glory of God. |

[C3] These several regions are again subdevided into various degrees or gradations of hapiness & misery according as they are situated nearer or farther remote from the extreems of either, but so sepparated by the Gulph of Mortality, that ye nature and laws to which either are subject are intirely unknown to ye other. The projection or plain of the picture you are to suppose parallel to ye *Orbis Magnus* and the perspective view penetrates so far into ye opposd concavity as to discover meriads of systems in all maner of dispostions with an infinite number of worlds [which] are variously distributed round their several Suns or Centers. And upon all sides the principal stars of the visible creation are exhibited in ther natural order as seen from ye Earth by ye naked eye. Those of ye first magnitude nearest to our own system, and the rest proportionable removed according to their resp[ec]tive phenomena. | [C4] Beyond these are others more remote crownd by a penumbral shaddow such as we call telescopic stars and again without them more, suposd to be at immense distance & by no means perceptible to ye human eye. At a certain distance from ye Sun equal to a vissual ray of ye smallest visible star is a faint circle of light terminating the utmost extent of ye visible creation, in a finite view from ye Earth; and all within this sphere is more or less inlightend by ye rays of our Sun according to their distance from him.

This vast creation of worlds I call the Gulph of Time or Region of Mortality, and next to this penetrating towards the abodes of the Blest you arive upon the Borders of Light, but if receeding in ye opposite direction you are led to Shades of Darkness & Desolation adjoining to ye presincts of the infinite abiss. |

[D1] The first mansion of immortality is imagind to be [the] place where ye Gardian Angels of ye human soul, present their charge to ye respective pations [?] of their faith. Here ye souls of ye Proffits, Patriarchs and

Primitive Fathers are first iniciated to ye Devine inspection and as subjects of Heaven obey ye sumons.

In this region as being ye first dawning sphere of immortal glory, the Royal Psalmest is represented playing upon his harp, praising God and his great creation; behind him is a gardian angell holding a wreath of palm over his head and an emblem of peace & endless joy; befor him is [D2] a lesser angel suporting his book of pious *Orisons.* | A little above upon one side is an angel resting upon a cloud, & supporting an orbicular simbol, representing ye eternal vicessitude or perpetual change of all created beings, and on ye other side is a long light vista leading to ye Primogenial Center of the Sedes Beatorum. At ye extremity of this luminous radia is ye figure of an eagle representing Contemplation, bearing in its claws a long scrawl or label with this inscription:

Strate is ye gate
and narrow is ye way,
that leadeth unto Life. |

[D3] On each side of ye great section supposd to be stationd in ye terrestrial regions are divers figures representing philosophers both ancient & modern, holding in their hands various schemes, demonstrating the several laws & principals of their respective relms explanatory of their motions, distance & phenomena. Here ye Geneous of Mathematicks is represented with a ballance, holding ye Sun & several planets in equilibrio and below is this motto:

He hangeth ye Earth upon nothing.

Opposite to this is ye Geneous of Geometry pointing to a diagram or construction, shewing the supernatural censure, catastrophy & conflagration of a world distroy'd. |

[D4] But now to reascend above ye Royal Psalmist our next step is into a more refined region and of still more exalted hapiness, in this sphere are suppos'd to be an infinite number of saints and angels, some of which are here represented in various postures of addoration; to whome a crown of glory is desending supported by two angelic boys each bearing the emblems of peacefull innocense along with ye regal rewards of vertue.

The next orb of Heaven approaches still nearer to perfection and is a relm of more & more refind beatitud. This I imagin to be a place of perfect joy & tranquili[t]y, full of the everlasting glories of ye Deaty, in that imortal state, which is next ye supreeme Heaven, all mutuale delite [E1] of consumate happiness & universal peace. | Here as upon his first admitence to the Devine Radience, is represented a transpo[r]ted Patriarch, struck with admiration and astonishment upon so visible an approximation to ye Devine Presence. He is neeling upon his knee unable to support himself under ye strong radient glories and superlative luster of the place. Here resting upon a cloud is an angel leaning upon a golden simbol, denoteing harmony & union. He is pointing towards ye Emperial Throne and over his head upon a flowing label is this sentence:

Glorious things are spoke of thee
O City of God. |

[E2] At some distance from this on ye opposite side seated upon an other cloud is another ang[el] represented as a scribe, he has a large book liing open before him in which is writ <this part of> ye sacred Te Deum or Bened:

We praise thee O God
We acknowledge thee to be the Lord.

Here, many angells with divers instruments of music are performing a celestial anthem to ye addoration of their devine presedents.

Within this circumambient region at various distance[s] from ye Center of Glory are many and vast groups of saints and angels, besides miriad [E3] other flying figures bearing various simbols of | peace, plenty, joy & love.

Allmost centering with perfection are infinite numbers of still more refind beings nearly uniteing with ye Deaty but of no determind image, seaming ye lest in glory.

The next and last step is to ye fountain of all perfection; the Center of Life & Immortality, but here ye imagination must conceive, what human language can't express, a place so full of the Divine Lust[re] that [E4] probably angels of ye 1t order, may only behold it with a coverd eye. | The Throne of Grace or of Devine Majesty, is expresst by ye following hyroglyptic:

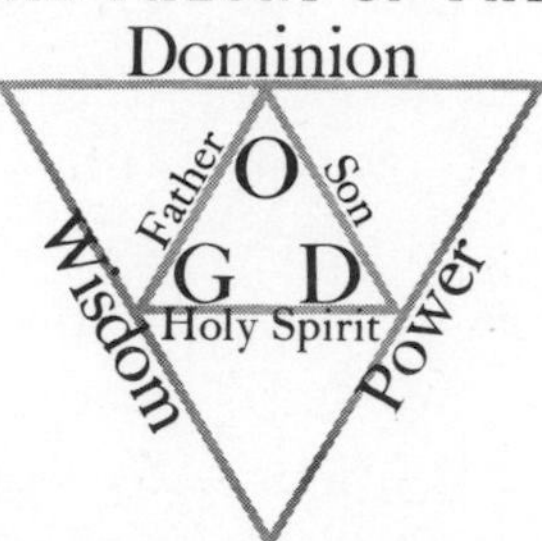

Right over this by which you are to understand the Emperial Seat, but at an angule distance from it, is a Book of the Holy Communion of Saints and Angels supported by two arch seraphims and in this to express [F1] in some measure the immeasurable concourse of the Heavenly | Host, with ye general joy and universal aclamations of angels upon the act of creation, is this verse:

> The Morning Stars sung together
> and all the Sons of God
> Shouted for joy.

Having now carried you thro' ye celestial regions I must now begin to receed and direct your attention towards ye infernal ones, but I am afrade I shall not be able to make the discription of them answer ye com̄on ideas. First I shall beg leave to suppose this rationaly, that if happiness increase as ye distance from perfection decreases, misery may also increase & torment as ye obje[c]ts are removed from the prime Presence of ye [F2] Deaty. | Consequently it may prove to be more than a conjecture, that those external regions beyond ye verge of this creation, may be ye commensment of an infinite abiss of desolation bordering either upon Chaos, or ye interstices of general night; this I would supose the place alloted for temporal punishments in a future state. In this dark region far remote from God, and void of all created being, Sattin and his angels are supposed to range in desolation, this being imagind the Horrid Mansion to which they were driven after their expulsion from ye Emperial Heaven. The inferior part of these eval agents are represented like snakes inter- [F3] woven with each other and grovelling in the utmost confusion. | Now as I don't know how to imagin that a spirit falling from a finite state thô in ye greatest excess of sinn can merit an equall degree of pain, with one fallen from a state of perfection, I have subdevided this Relm of Darkness into three different states, and that bordering upon the Eternal Night I have given to the most demonical spirits, the primary leaders or inspirators of the primitive rebelion such as the Devil and his original revol[t?]ing Agents. Here ye great arch angel Michael, in all his radient glories, is represented brandishing the thunder of God upon ye arch offende[r] | [F4] which appears of a gloomy aspect under his feet, attended by Envy in a like disabled posture teasing her snakes. Upon Michaels sheeld is inscribed this word

Invinsible.

Bordering upon this region towards the Gulph of Time is a much milder sphere where you are to imagin the secondary angels or procelites who fell without premeditated designe, more by ye ill advice and influence of others than self choruptors. This region is faintly inlightened by a dubious [G1] dawn proceding from a confused | light reflected from the luminous bodies in ye Region of Mortality or Relms of Death. Here Sin & Death are represented preparing an invasion against ye Human Species and at some distance is a weeping boy struck with a sense of future ills and is lamenting ye state of Man, being subject to mortality & obnoxions on all sides [and] to ye insults of Hell. By him is this line:

Wo to ye inhabitants on Earth. |

[G2] Next above these joining to ye verge of creation is ye state appointed for ye human souls or ye punishment of mortals. Here the eye may be supposed able to penetrate into ye abiss of motion, as in some part of the abiss it self the human eye <in> its mortal state may perhaps se[e] some of ye eternal beams proceeding from ye Seat of Glory. In this case ye wicked thô in a state of pain may possibly behold on one side a bright and glorious creation, and on ye other nothing but gloomy darkness & horrid obscurity.

Here as a seene of punishment is represented a human figure chained upon a rock of cediment [?] to a penal bed of undisolving icce, & near him is written this: |

This is the state of a wicked man.

[G3] We may suppose that after our mortal change the souls of [the] wicked do not imediately enter upon bodily paine or any kind of material punishment, but as in a state of oblivion are prepar'd by painfull visions or tormenting dreames arising by a kind of internal[?] reflection from ye near approaching scene of Wo in order to <strengthen ye mind &> to enable them to support such dreaded pains. In like manner we may imagine the <souls of ye> good return to rest after death and find an Elizium in the grave and anticipate ye joys to come. |

[G4] The situation of these several regions of joy & pain in a very lively manner p[rese]nts to our imagination that misterious parable of *Dives & Lazarus* where ye Patron of Christianity endeavouring to give the Jews and Gentiles, not only an idea of ye weakness of human faith in this life but also an imaginary view of ye fatal distance of Heaven & Hell in a future life, tells them by way [of] allegory in ye person of Abram that those two places are so situated and at so vast a distance from each other [H1] that betwixt them ly an impassable gulph impenetrable on | either side, but by ye souls of ye dead who by a vertuous or impious death are permitted to pass from this world to happiness or missery in the next. This image of futurity I have emblematically represented consentanious to Scripture by ye three different states, viz: the present problematical, ye future good & ye future ill.

At ye lower end of ye scheme are placd two erect figures as spectators of ye whole creation and above one of them is this verse:

> The visible things of God, from the
> creation of ye World is clearly seen
> being understood by ye things that
> are made even his eternal power &
> godhead.
>
> Rom 1 : 20th. |

[H2] and over ye other this.

> The Heavens declare ye glory of God
> and the firmament sheweth his
> handy work.
>
> Psalms XIX 1.

In the middle betwixt ye fore mentiond figures is ye arch angel Gabriel treading upon a clood and pointing to ye lower parts of ye scheme and over his head are these laws.

> Stand still and consider the wondrous works of God.
> Job : XXXVII, 14.
> He is far from ye wicked. Prov. XV. 29.
> Behold what desolation he has made.

Lastly ye scheme closes with two figures representing Chaos and Old Night. |

[H3] <This juvenil performance was the produce of the authors imagination before he had reap'd any advantages either from reading or study, but prov'd afterwards the foundation of his Theory of y^e Univers a much more perfect work.>

AN

ORIGINAL THEORY

OR

NEW HYPOTHESIS

OF THE

UNIVERSE,

Founded upon the

LAWS of NATURE,

AND SOLVING BY

MATHEMATICAL PRINCIPLES

THE

General PHÆNOMENA of the VISIBLE CREATION;

AND PARTICULARLY

The VIA LACTEA.

Compris'd in Nine Familiar LETTERS from the AUTHOR to his FRIEND.

And Illustrated with upwards of Thirty Graven and Mezzotinto Plates, By the Best MASTERS.

By THOMAS WRIGHT, of DURHAM.

One *Sun by Day, by Night* ten Thousand *shine,*
And light us deep into the DEITY. Dr. YOUNG.

LONDON:
Printed for the AUTHOR, and sold by H. CHAPELLE, in *Grosvenor-Street.*
MDCCL.

THE

PREFACE.

THE Author of the following Letters having been flattered into a Belief, that they may probably prove of ſome Uſe, or at leaſt Amuſement to the World, he has ventured to give them, at the Requeſt of his Friends, to the Publick. His chief Deſign will be found an Attempt towards ſolving the Phænomena of the *Via Lactea*, and in conſequence of that Solution, the framing of a regular and rational Theory of the known Univerſe, before unattempted by any. But he is very ſenſible how difficult a Taſk it is to advance any new Doctrine with Succeſs, thoſe who have hitherto attempted to propagate aſtronomical Diſcoveries in all Ages, have been but ill rewarded for their Labours, tho' finally they have proved of the greateſt Benefit and Advantage to Mankind. This ungrateful Leſſon we learn from the Fate of thoſe ingenious Men, who, in ignorant Times, have unjuſtly ſuffered for their ſuperior Knowledge and Diſcoveries; they who firſt conceived the Earth a Ball, were treated only with Contempt for their idle and ridiculous Suppoſition, as it was called; and he who firſt attempted to explain the *Antipodes*, loſt his Life by it; but in this Age Philoſophers have nothing to fear of this ſort, the great Diſadvantages attending Authors now, are of a widely different

A 2 Nature,

Nature, rising from the infinite Number of Pretenders to Knowledge in this Science, and much is to be apprehended from improper Judges, tho' from real ones nothing; for nothing is more certain than this, as much as any Subject exceeds the common Capacity of Readers, so much will the Work in general be condemned; the Air of Knowledge is at least in finding Fault, and this vain Pretence generally leads People, who have no real Foundation for their Judgment to argue from, to ridicule what they are too sensible they do not understand. Thus the same Disadvantages too often attend both in publick and private an exceeding good Production equally the same as a very bad one: But the Author is not vain enough to think this Work without Faults, has rather Reason to fear, from the Weakness of his own Capacity, that there may be many; but he hopes the Design of the Whole will, in some measure, plead for the Imperfection of the Parts, if the Merits of the Plan should be found insufficient for his full Pardon, in attempting so extensive a Subject.

In a System thus naturally tending to propagate the Principles of Virtue, and vindicate the Laws of Providence, we may indeed say too little, but cannot surely say too much; and to make any further Apology for a Work of such Nature, where the Glory of the Divine Being of course must be the principal Object in View, would be too like rendering Virtue accountable to Vice for any Author to expect to benefit by such Excuse. The Motive which induces us to the Attempt of any Performance, where no good Reason can be supposed to be given for the Omission, or Neglect of it, will always be judged an unnecessary Promulgation, and consequently every Attempt towards the Discovery of Truth, the Enlargement of our Minds, and the Improvement of our Understandings will naturally become a Duty. If therefore this Undertaking falls short of being instrumental towards the advancing the Adoration of the Divine Being in his infinite Creation of higher Works, and proves unable to answer all Objections that may possibly arise against it, yet will its Imperfections appear of such a Nature to every candid Reader, as to afford the Author a sufficient Apology for producing them to the World: And it is to be hoped farther, that where a Work is entirely upon a new Plan, and the Beginning, as it were, of a new Science, before unattempted in any Language, the Author having dug all his Ideas from the Mines of Nature, is surely intitled to every kind of Indulgence.

To

To thoſe who are weak enough to think that ſuch Enquiries as theſe are over-curious, vain, and preſumptive, and would willingly, ſuitable to their own Ignorance and Comprehenſion, ſet Bounds to other People's Labours, I anſwer with Mr. *Huygens*, " That if our Forefathers had " been at this Rate ſcrupulous, we might have been ignorant ſtill of the " Magnitude and Figure of the Earth; or that there was ſuch a Place as " *America*. We ſhould not have known that the Moon is enlightened by " the Sun's Rays, nor what the Cauſes of the Eclipſes of each of them " are; nor a Multitude of other Things brought to Light by the late " Diſcoveries in Aſtronomy; for what can a Man imagine more abſtruſe, " or leſs likely to be known, than what is now as clear as the Sun."

Had we ſtill paid that Homage to a Name,
Which only God and Nature juſtly claim;
The weſtern Seas had been our utmoſt Bound,
Where Poets ſtill might dream the Sun was drown'd;
And all the Stars that ſhine in Southern Skies,
Had been admir'd by none but ſavage Eyes.

DRYDEN.

Beſides the Nobleneſs and Pleaſure of theſe Studies, *Wiſdom* and *Morality* are naturally advanced, and much benefited by them, and even Religion itſelf receives a double Luſtre, " to the Confuſion of thoſe who " would have the Earth, and all Things formed by the ſhuffling Concourſe " of Atoms, or to be without Beginning." In Aſtronomy, as well as in natural Philoſophy, though we cannot poſitively affirm every thing we ſay to be Facts and Truth, yet in ſo noble and ſublime a Study as that of *Nature*, it is glorious, as Mr. *Huygens* ſays, even to arrive at Probability.

Notwithſtanding then the Diſadvantages which ever have attended all new Diſcoveries, either thro' the Ignorance of the Age, or the univerſal Paſſion of Ridicule in ſuch contented Creatures, as can't comprehend, yet ever attacking with a fool-hardy Reſolution, the advancing Enſigns of Knowledge, if Ignorance was Virtue, and Wiſdom Vice; I ſay, regardleſs of this noiſy Shore, it is ſure our Duty to ſpring forward, and explore the ſecret Depths of Infinity, and the wonderful hidden Truths of this vaſt Ocean of Beings. But how the heavenly Bodies were made, when they were made

made, and what they are made of, and many other Things relating to their Entity, Nature, and Utility, ſeems in our preſent State not to be within the Reach of human Philoſophy; but then that they do exiſt, have final Cauſes, and were ordained for ſome wiſe End, is evident beyond a Doubt, and in this Light moſt worthy of our Contemplation.

He who thro' vaſt Immenſity can pierce,
See Worlds on Worlds compoſe one Univerſe,
Obſerve how Syſtem into Syſtem runs,
What other Planets, and what other Suns;
What varied Being peoples ev'ry Star;
May tell why Heav'n made all Things as they are.

POPE.

To expect that ſo new an Hypotheſis ſhould meet with univerſal Approbation, would be an unpardonable Vanity; nor is it reaſonable every Reader ſhould think the Author obliged to remove all his Prejudices and Partialities, ſo far as to give him the perfect Picture of the Univerſe he likes beſt. In many Caſes it would be ſo far from being better for the World, if all Men judged and thought alike, that Providence ſeems rather to have guarded againſt it as an Evil, than any how to have promoted it as a general Good: But the following Theory regards the Whole rather than Individuals: And the many worthy Authors cited in the Work, who have all greatly favoured this extenſive Way of Thinking, will, I hope, be a ſufficient Excuſe for forming theſe obvious Conjectures into a Theory, eſpecially where ſo great a Problem is attempted as the Solution of the *Via Lacteal* Phænomenon, which has hitherto been looked upon as an inſurmountable Difficulty. How the Author has ſucceeded in this Point, is a Queſtion of no great Conſequence; he has certainly done his beſt; another, no Doubt, will do better, and a third perhaps, by ſome more rational Hypotheſis, may perfect this Theory, and reduce the Whole to infallible Demonſtration: The firſt Syſtem of the ſolar Planets was far from a true one, but it led the Way to Perfection, and the laſt we can never too much admire. It is well known, that the firſt Syſtem of the Planets was alſo but a Conjecture, yet none will deny that it was an happy one.

The

The Discovery of the Magnet Poles; the Government of the Tides; proportional Distance and Periods of the Planets, &c. have all their Uses, and undoubtedly were designed to be known. Ignorance is the Disgrace of Mankind, and sinks human Nature almost to that of Reptiles. Knowledge is its Glory and the distinguishing Characteristic of rational Creatures.

To Enquiries of this sort, then sure we may say with *Milton*, That

GOD'S OWN EAR LISTENS DELIGHTED.

The Subject is, no Doubt, the noblest in Nature, and as such, will always merit the Attention of the thinking Part of Mankind. Men of Learning and Science, in all Ages, have ever made it their peculiar Study. Towards the latter End of the Republic, and afterwards in the more peaceable Times of *Trajan* and the *Plinys*, we have no Reason to doubt but that Astronomy was in the highest Reputation: And notwithstanding *Greece* had been the chief Seat of the Philosophers, yet may we suppose *Rome* in those Days little inferior in the Knowledge of the Stars, when we find Men * of the first Figure in Life become Authors upon the Subject.

We have many Instances to shew, that Astronomy was in the greatest Repute amongst the Antients of all Ranks, and almost every where looked upon as one of the greatest, if not as one of the first Qualifications of their best Men. As a Confirmation of which, we find in the historical Accounts of the *Argives*, a very warm Contest betwixt the two Sons of *Pelops* 1205 Years before *Christ*, thus testified by *Lucian:* When the *Argives*, by publick Consent, had decreed that the Kingdom should fall to him of the two, who should manifest himself the most learned in the Knowledge of the Stars, *Thyestes* thereupon is said to have made known to them, the Constellation, or Sign of the *Zodiack* call'd *Aries:* But *Atreus* at the same time discovering to them the Course of the Sun, with his various Rising and Setting, demonstrating his Motion to be * contrary to that of the Heavens, or diurnal Motion of the Stars, was thereupon elected King.

* *Cicero* translated the Phænomena of *Aratus* into *Latin* Verse. *Julius Cæsar*, as *Pliny* relates, wrote of Astronomy in *Greek*, and is said to have left several Books of the Motion of the Stars behind him, derived from the Doctrine of the *Egyptians*. *Ant. Chris.* 45. He with *Sosigenes* reformed the *Roman* Year, which was first invented by *Numa Pompilius*. *Germanicus Cæsar* also translated *Aratus*'s Phænomena into *Latin* Verse *Anno Dom.* 15. *Tiberius* and *Hadrian* are also said to have wrote on Astronomy.

* Hence arose the Fable of the Sun's going backwards in the Days of *Atreus*, as if struck with Abhorrence of his bloody Banquet. *Vide Ovid*'s Metamorphosis.

To

To recite more of the moſt eminent Patrons and Profeſſors of this kind of Learning here, will carry me too far from my preſent Purpoſe; for farther Information therefore, I ſhall refer the inquiſitive Reader, to that curious Catalogue in *Sherburn*'s Sphere of *Manilius*, where ſo many ruling † Men of all Ages and Nations ſwell, and illuſtrate the Number.

In a Word, when we look upon the Univerſe as a vaſt Infinity of Worlds, acted upon by an eternal Agent, and crouded full of Beings, all tending through their various States to a final Perfection, and reflect upon the many illuſtrious Perſonages, who have, from time to time, thought it a kind of Duty to become Obſervers, and conſequently Admirers of this ſtupendious Sphere of primary Bodies, and diligent Enquirers into the general Laws and Principles of Nature, who can avoid being filled with a kind of enthuſiaſtic Ambition, to be acknowledged one of the Number, who, as it were, by thus adding his Atom to the Whole, humbly endeavours to contribute towards the due Adoration of its great and divine Author.

I judge it will be quite unneceſſary to ſay any thing about the Order of the Work, ſince that would be only a Repetition of the Table of Contents, to which the Reader is referred, as to the propereſt Account that can here be given.

† Seven Emperors, nine Kings, and as many ſovereign Princes. *Charlemagne* wrote *Ephemerides*, and named the Months and Winds in *High Dutch*, 770. *Rich.* II. &c.

THE

THE

CONTENTS.

DIRECTIONS for placing the PLATES.

Plate.	*Page.*	*Plate.*	*Page.*	*Plate.*	*Page.*	*Plate.*	*Page.*
I.	10	IX.	22	XVII.	51	XXV.	64
II.	11	X.	23	XVIII.	52	XXVI.	*ibid.*
III.	16	XI.	35	XIX.	56	XXVII.	64
IV.	20	XII.	38	XX.	*ibid.*	XXVIII.	65
V.	*ibid.*	XIII.	*ibid.*	XXI.	62	XXIX.	*ibid.*
VI.	22	XIV.	40	XXII.	63	XXX.	70
VII.	*ibid.*	XV.	42	XXIII.	64	XXXI.	83
VIII.	*ibid.*	XVI.	*ibid.*	XXIV.	*ibid.*	XXXII.	*ibid.*

Some of the Principal ERRATA.

Page	*Line*	*the Words*	*Read.*
2	*ult.*	to ceaſe relating	ceaſing to relate
4	3	Phænomenon	Phænomena
16	15	incomſible	incomprehenſible
21	12	comprehend	comprehending
33	28	compared	is compared
34	37	form	from
43	20	volving	revolving
49	24	immoveable	moveable
61	19	much	much as
62	28	XXIII.	XXI.
65	4	where	any where
67	15	alſo	all ſo
69	29	one	our

Plate X. read the Characters of the Planets in this Order ♃ ☿ ♄ ♂ ♀

THE

A

L I S T

OF THE

SUBSCRIBERS.

A.

LORD Anſon.
Hon. Mr. Archer.
Charles Ambler, Eſq;

B.

Duke of Beaufort.
Duke of Bedford.
Dutcheſs of Beaufort.
Lord Berkely, *of* Straton.
Miles Barne, *Eſq;*
Lancelot Barton, *Eſq;*
Hon. Antoine Bentinck.
Hon. John Bentinck.
Norbone Berkely, *Eſq;*
John Brown, *Eſq;*
—— Blaman, *Eſq;*
Thomas Brand, *Eſq;*
J. Bevis, *M. D.*
Rev. T. Bonney, *A. M.*

C.

Counteſs of Cunengeſby.
Lord Cornwallis.
Lady Cornwallis.
Edward Cave, *Eſq;*
John Chamock, *Eſq;*
Hon. and Rev. Dr. Cowper.
Mr. Richard Chad.
Mr. Henry Chapell.
Iſ. Colepepper.
Mr. George Conyers.

D.

Rev. John Dealtary, *A. M.*
Mr. Samuel Dent.

F.

Charles Fitzrea Scudamore, *Eſq;*
Kean Fitzgerald, *Eſq;*
Thomas Fonnerau, *Eſq;*
Robert Rakes Fulthorpe, *Eſq;*
Mr. Samuel Farrant.
Mr. Paul Fourdrinier.

G.

Marchioneſs Grey.
Lord Glenorchy.
Francis Godolphin, *Eſq;*
Roger Gale, *Eſq;*
James Gibbon, *Eſq;*
Ralph Goward, Eſq;
Ralph Gowland, *Eſq;*
Ralph Gowland, *Junior*, *Eſq;*
Dr. Gregory.
Dr. Griffith.
Rev. John Griffith, *A. M.*
Rev. Middlemore Griffith.

H.

Lord Hardwick, *Lord High Chancellor of* Great-Britain.
Hon. James Hamilton.
Mr. Thomas Heath.
Mr. Thomas Holt.
John Hughes, *Eſq;*

Earl

A LIST *of the* SUBSCRIBERS.

I.

Earl of Jerſey.
Richard Jackſon, *Eſq;*
Rev. Mr. Jones.

K.

——— Knowles, *Eſq;*
Dr. Kendrick.
Mrs. Kennon, 4.

L.

Lady Vicounteſs Limerick.
Sir William Lee, *Bart.*
William Leſter, *Eſq;*
Rev. Dr. Long, *Maſter of* Pembroke-hall, Cambridge.
William Lloyd, *Eſq;*
Mr. Andrew Lawrence.

M.

R. J. Mead, *M. D.*
Richard Meyrick, *M. D.*
Owen Meyrick, *Eſq;*
Pierce Meyrick, *Eſq;*

N.

Duke of Norfolk.
Lord North.
Lord Biſhop of Norwich.
Richard Nicholls, *Eſq;*
Mrs. Norſa.

P.

Duke of Portland.
Earl of Pembroke, *&c.* 2
Counteſs of Pembroke, *&c.*
Lady Palmerſton.
Robert Money Penny, *Eſq;*
Sir Francis Pool.
Sir John Pool.
John Probyn, *Eſq;*
Rev. Mr. Pierce.
Mr. Dominick Pile.
Mr. Powel, *of* Cambridge,

R.

Dutcheſs of Richmond, *&c. &c.*
James Ralph, *Eſq;*
Allan Ramſey, *Eſq;*
William Read, *Eſq;* 2.
Henry Reveley, *Eſq;*
William Reveley, *Eſq;*

S.

Sir George Savile.
——— Serle, *Eſq;*
Rev. Dr. Smith, *Maſter of* Trinity College, Cambridge,
Miſs Stonehouſe.
William Symonds, *Eſq;*
Mr. James Scot.
Mr. James Stephens.

T.

Lord Viſcount Townſhend.
John Temple, *Eſq;*
James Theobald, *Eſq;*
Charles Townſhend, *Eſq;*
Mrs. Mary Trevor.
Mr. James Thornton.

V.

Lord Viſcount Villiers.

W.

Lady Frances Williams.
Miſs Williams.
Miſs Charlotta Williams.
Rev. Thomas White, *A. M.*
——— White, *Eſq;*
Charles Louis Wiedmarkter, *Eſq;*
Mr. Ward.

Y.

Hon. Philip York.
Dr. Arthur Young, *Preb. of* Cant.

LETTER THE FIRST.

Opinions of the most eminent Authors whose Sentiments on the following Subject have been published in their Works.

SIR,

EFLECTING upon the agreeable Converſation of our laſt Meeting, which you may remember chiefly turned upon the Stars, and the Nature of the planetary Bodies; a Subject, which is generally allowed to give true Pleaſure to all thoſe who take Delight in mathematical Enquiries; and having not a little Regard to the repeated Requeſt in your late Letters, I have at length undertaken to explain to you, as far as I am able, my Theory of the *Univerſe*, and the Ideas I have form'd of the known Creation.

The Hypotheſis upon which this new Aſtronomy is founded, and now reduced into a regular Syſtem, was the reſult of my Aſtronomical Studies * full fifteen years ago, hence I hope you will allow, I have more than obſerved *Horace*'s celebrated Aphoriſm,

Nonumque prematur in annum.

* The firſt Scheme of this Hypotheſis was plann'd in the Year 1734, repreſenting in a Section of the Creation, eighteen Feet long and one broad, ſeveral thouſand Worlds and Syſtems, and a great Number of emblematical Figures, now in the Author's Poſſeſſion, together with a Scheme of the entire Creation, completed ſince, nine Feet long and ſix broad, more fully illuſtrating upon the ſame Conſtruction the Innumerability of Syſtems and Worlds.

The Subject, I have often observed, you have listened to with a pleased Attention, and I am the more incouraged to explain it at large to you, as I am perswaded you don't want to be convinced of its valuable Uses and Importance.

I remember you have often told me, that to apply ourselves to the Study of Nature, was the surest and readiest Way to come at any tolerable Knowledge of ourselves, however difficult the Task might prove either in the Attempt, or the attaining it, and the less to be neglected, as it never fails to introduce a proper Knowledge of the DIVINE BEING, as a certain Consequence along with it, and such a Knowledge, as will naturally make every Man, who has but a tolerable Share of common Sense, and is not a Slave to another's Reason, without any other Evidence or Motive, in all Stations, and under all Circumstances, ACT JUSTLY, LIVE CHEARFULLY, and DIE full of Hope in the Expectation of a happy Sequel, in Futurity.

> *Eternity* is written in the Skies:
> Mankind's Eternity, nor *Faith* alone;
> *Virtue* grows there ———
>
> *Dr.* YOUNG.

A learned Author on the Attributes, recommending these Studies as a reasonable and moral Service, says, " Sure, it is most becoming such im-" perfect Creatures as we are, to contemplate the Works of God with this " Design, that we may discern the Manifestations of Wisdom in them; " and thereby excite in ourselves those devout Affections, and that su-" perlative Respect, which is the very Essence of Praise."

> Who turns his Eye, *on Nature's Midnight Face*,
> *But must enquire* ——— what Hand behind the Scene,
> What ARM ALMIGHTY, put these wheeling Globes
> In Motion, and wound up the vast Machine?

The enchanting Idea *Milton* had of the Subjects of Astronomy (whose truly sublime Way of thinking and writing perhaps was never so nearly equalled, or attempted before this Reverend Author's *Night-Thoughts*, appear'd is finely shewn in the Eighth Book of his *Paradise Lost*, where he makes his *Adam*, so earnestly attentive to the Angel *Gabriel*, as to cease relating the Mysteries of Creation.

Raphael

The

The Angel ended, and his *Adam*'s Ear
So charming left his Voice, that he awhile
Thought him ſtill ſpeaking; ſtill ſtood fix'd to hear.

Milton's own Ideas of the Univerſe too, which no doubt he had gathered from aſtronomical Authors, and had reconciled himſelf to, we are fully made acquainted with in the ſame Book, where the Arch-angel ſays, in anſwer to *Adam*'s Enquiries.

—— Other Suns perhaps
With their attendant Moons thou wilt deſcry
Communicating Male and Female Light,
Which two great Sexes animate the World,
Stor'd in each Orb, perhaps with ſome that live:
For ſuch vaſt Room in Nature, unpoſſeſt
By living Soul, deſert and deſolate,
Only to ſhine, yet ſcarce to contribute
Each Orb a Glimpſe of Light, convey'd ſo far
Down to this habitable, which returns
Light back to them, is obvious to Diſpute.

But before I preſume to plan my own Diſcoveries and Conjectures into a Theory, both in Juſtice to thoſe who have in ſome meaſure been in the ſame Way of Thinking, and alſo as a Defence of myſelf for producing ſo new an Hypotheſis to the World, which otherwiſe (though any Apology made to you I know will be unneceſſary) may appear to too many but an idle *Chimera* of my own. I judge it will be highly proper, by way of ſtrengthening my own Arguments, and adding more Weight to what I ſhall myſelf advance in the following Letters, to give you in this the Opinions of the moſt able Writers, whoſe Works I have read upon the Subject. I mean ſo far as relates to the now general received Notion, that the Stars are all Suns, and ſurrounded with planetary Bodies, with which I ſhall ſet out; and ſhew you, it is not a Thing merely taken for granted, but has ever been the concurrent Notion of the Learned of all Nations, as ſhall be further ſhewn, in its proper Place, and as nearly as Poſſibility will admit of, demonſtrated to be Truth.

The following is an Extract from Mr. *Toland*, in his Account of the Works of

JORDANUS BRUNO.

" The Divine Efficacy (ſays this Author in his infinite Creation) cannot " ſtand idle, without the Want of Will or Power; but any Imbecillity in

" ſuch a Being argues Imperfection, and ſince any finite Produce compared with Infinity is as nothing, or rather as the Beginning of Good, " it muſtbe no leſs idle, and invidious in producing a finite Effect, than in " producing none at all.

" Hence, as all Finites, ſingly conſidered, are but as Commencements " of ſomething more to be expected.

" Omnipotence, in making the Creation finite, will appear to be no " leſs blameable for not being willing, than for not being able, to make it " otherwiſe; *i. e.* infinite, as being an infinite Agent upon a finite Subject, " which is repugnant to Reaſon."

It follows then that, Creation muſt be not only extenſively, but intenſively indefinite, and beyond the Reach of the human Underſtanding to comprehend; and that the one is as neceſſary as the other, *i. e.* an infinite Expanſe is as reconcileable to our Reaſon, as infinite Parts are to our Senſes.

All the Attributes of the Divine Being are, as any one of them, incomſible to his Creatures; why ſhould our Imagination then be ſuppoſed to extend beyond the divine Activity?

" Thus, adds the above Author, the Excellency of God is adequately " magnified, and the Grandeur of his Empire made manifeſt; he is not " glorified in one, but in numberleſs Suns; not in one Earth, or in one " World, but in ten thouſand thouſand of infinite Globes."

An infinite Repreſentation of an infinite Original, and a Spectacle befitting the Excellency and Eminence of him, that can neither be fully conceived, imagined, or comprehended.

> What read we here? th'Exiſtence of a God?
> Yes, and of other Beings, Man above,
> Natives of Æther! Sons of higher Climes!
>
> Dr. Young.

" If the Exiſtence of this one World be good or convenient, it is not " leſs good or convenient that there be infinite others like it.

" The infinite efficient Cauſe would be abſolutely defective, without an " infinite Effect; and beſides, by conceiving the Infinity of the Univerſe " and innumerable Beings, the Underſtanding reſts ſatisfied, and is reconciled with the Idea of an Eternity; whereas, by aſſerting the contrary, " it is unavoidably plunged into innumerable Difficulties, and unſolvable " Inconveniencies, Paradoxes, and Abſurdities.

Again, ſays the ſame Writer, " Did we but conſider and comprehend " all this, oh! to what much further Conſiderations and Comprehenſions " ſhoud

" ſhould we be carried! as we might be ſure to obtain that Happineſs
" by virtue of this Science, which *in other Sciences is ſought after in vain.*

> This Proſpect vaſt, what is it? weigh'd aright,
> 'Tis Nature's Syſtem of Divinity,
> And every Student of the Night inſpires.
>
> Dr. YOUNG.

> 'Tis elder Scripture, writ by GOD's own Hand;
> Scripture authentic! uncorrupt by Man.

" This then is that Philoſophy, which opens the Senſes, which ſatisfies
" the Mind, which enlarges the Underſtanding, and which leads Man-
" kind to the only true Beatitude, whereof they are capable according to
" their natural State and Conſtitution; for it frees us from the ſollicitous
" Purſuit of Pleaſure, and from the anxious Apprehenſions of Pain, mak-
" ing us to enjoy the good Things of the preſent Hour, and not to fear
" more, than we hope from the future; ſince that ſame Providence, or
" Fate, or Fortune, which cauſes the Viciſſitudes of our particular Being,
" will not let us know more of the one, than we are ignorant of the
" other."

And farther, " From theſe Contemplations, if we do but rightly conſider,
" it will follow, that we ought never to be diſpirited by any ſtrange Ac-
" cidents, through Exceſs of Fear or Pain, nor ever be elated by any proſ-
" perous Event, through Exceſs of Hope or Pleaſure; whence we have
" the Path to true Morality, and following it, we ſhall of courſe become
" the magnanimous Deſpiſers of what Men of weak Minds fondly
" Eſteem, and be wiſe Judges of the Hiſtory of Nature, which would be
" written in our Minds, and conſequently be chearful and ſtrict Execu-
" tioners of the divine Laws, which would thus be ingraved in the Cen-
" ter of our Hearts. Seeking, as it were, in ourſelves, an Approbation of
" our own Action, which alone is capable of true Content and Happi-
" neſs."

CHRISTOPHER HUYGENS,

To whom the World is much indebted for many curious Inventions, and Diſcoveries, ſays in his *Planetary Worlds*, " I muſt be of the ſame
" Opinion with all the great Philoſophers of our Age, that the
" Sun is of the ſame Nature with the fix'd Stars; and this will give us a
" greater

† The Pendulum Clock; the firſt Diſcovery of *Jupiter*'s Satellites, and *Saturn*'s Ring.

" greater Idea of the World than all other Opinions can. For then
" why may not every one of theſe Stars, or Suns, have as great a Retinue,
" as our Sun, of Planets, with their Moons to wait upon them? Nay,
" there is a manifeſt Reaſon why they ſhould; for, if we imagine our-
" ſelves placed at an equal Diſtance from the Sun and fix'd Stars, we
" ſhould then perceive no Difference at all betwixt them.

" Why then may we not make uſe of the ſame Judgment that we
" would in that Caſe; and conclude, that our Star has no better Atten-
" dance than the others? So that what we allowed the Planets upon the
" Account of our enjoying it, we muſt likewiſe grant to all thoſe Planets
" that ſurround that prodigious Number of Suns. They muſt have their
" Plants and Animals, nay, their rational Creatures too, and thoſe as great
" Admirers and as diligent Obſervers of the Heavens as ourſelves; and
" muſt conſequently enjoy whatever is ſubſervient to, and requiſite for
" ſuch Knowledge.

" What a wonderful and amazing Scheme have we here of the mag-
" nificent Vaſtneſs of the Univerſe! So many Suns, ſo many Earths, and
" every one of them ſtock'd with ſo many Herbs, Trees, and Animals,
" and adorned with ſo many Seas and Mountains! And how muſt our
" Wonder and Admiration be increaſed, when we conſider the prodi-
" gious Diſtance and Multitude of the Stars?"

The Opinion of Sir ISAAC NEWTON.

This great Author, in his grand *Scholia* to the *Principia*, ſays: --- " The
" moſt beautiful Syſtem of the Sun, Planets, and Comets, could only pro-
" ceed from the Counſel and Dominion of an intelligent and powerful
" Being: And if the fix'd Stars are the Centers of other like Syſtems, theſe,
" being form'd by the like wiſe Counſel, muſt be all ſubject to the Do-
" minion of One; eſpecially, ſince the Light of the fix'd Stars is of the
" ſame Nature with the Light of the Sun, and from every Syſtem Light
" paſſes into all the other Syſtems. And leaſt the Syſtems of the fix'd
" Stars ſhould by their Gravity fall mutually on each other, he (the Di-
" vine Being) hath placed thoſe Syſtems at immenſe Diſtances from one
" another."

The

The Opinion of Dr. DERHAM, *in his* Aſtro-Theology.

" The new Syſtem, ſays he, ſuppoſeth there are many other Syſ-
" tems of Suns and Planets, beſides that, in which we have our
" Reſidence; namely, that every fix'd Star is a Sun, and incompaſſed
" with a Syſtem of Planets, both primary and ſecondary, as well as ours.
" Theſe ſeveral Syſtems of the fixed Stars, as they are at a great and
" ſufficient Diſtance from the Sun and us; ſo they are imagined to be at
" as due, and regular Diſtances from one another: By which means it is
" that thoſe Multitudes of fixed Stars appear to us of different Magnitudes,
" the neareſt to us large; thoſe farther and farther, leſs and leſs; and
" that ſome, if not all of thoſe vaſt Globes of the Univerſe, have a Mo-
" tion, is manifeſt to our Sight, and may eaſily be concluded of all, from
" the conſtant Similitude and Conſent that the Works of Nature have
" with one another."

To this we may add, that this Syſtem of the Univerſe, as it is phyſi-cally demonſtrable, is far the moſt rational and probable of any. *Becauſe*,
" It is far the moſt magnificent of any, and worthy of an infinite
" CREATOR, whoſe *Power* and *Wiſdom*, as they are without Bounds and
" Meaſure, ſo may they, in all Probability, exert themſelves in the Creation
" of many Syſtems as well as one. And as Myriads of Syſtems are more
" for the *Glory* of GOD, and more demonſtrate his *Attributes* than one;
" ſo it is no leſs probable than poſſible, there may be many beſides this
" which we have the Privilege of living in." And as the ſtrongeſt Con-firmation of this, " we ſee it is really ſo, as far as it is poſſible it can be
" diſcerned by us, at ſuch immenſe Diſtances as thoſe Syſtems of the fixed
" Stars are from us; and we cannot reaſonably expect more."

" Since the Sun and fix'd Stars, ſays Dr. *Gregory*, are the only great
" Bodies of the Univerſe that have any native Light, they are juſtly
" eſteemed by Philoſophers to be of the ſame Kind, and deſigned for the
" ſame Uſes; and it is the Effect of a Man's Temper that ſets a greater
" Value upon his own Things than he ought, that makes him judge
" the Sun to be the biggeſt of them all."

That, as an elegant * Writer obſerves, which we call the Morning, or the Evening Star, is, in reality, a *Planetary World*; which, with the four others, that ſo wonderfully, as *Milton* expreſſes it, " vary their myſtick
" Dance, are in themſelves dark Bodies, and ſhine only by Reflection;
" have Fields and Seas, and Skies of their own; are furniſhed with all
" Accommodations for animal Subſiſtence, and are ſuppoſed to be the
Abodes

cf. 8. 148

cf. 5. 616ff

* Contemplations on the ſtarry Heavens.

" Abodes of intellectual Life. Again, The Sun, with all its attendent Planets
" is but a very little Part of the grand Machine of the Univerſe. Every
" Star— is really a vaſt Globe, like the Sun, in Size and in Glory, no leſs
" ſpacious, no leſs luminous, than the radiant Source of our Day; ſo that
" every Star is the Center of a magnificent Syſtem, has a Retinue of
" Worlds irradiated by its Beams, and revolves round its active Influence;
" all which are loſt to our Sight in immeaſurable Tracts of Æther.

" Could we, ſays the ſame Author, wing our Way to the higheſt ap-
" parent Star — we ſhould there ſee other Skies expanded, other Suns,
" that diſtribute their inexhauſtible Beams of Day; other Stars, that gild
" the alternate Night; and other perhaps nobler Syſtems eſtabliſhed;
" eſtabliſhed in unknown Profuſion, through the boundleſs Dimenſions
" of Space. Nor does the Dominion of the great Sovereign end *there*,
" even at the End of this vaſt Tour, we ſhould find ourſelves advanced
" no farther than the Frontiers of Creation; arrived only at the Suburbs
" of the great *Jehovah*'s Kingdom."

cf 8. 348

O for a Teleſcope his Throne to reach!
Tell me ye Learn'd on Earth! or Bleſt above!
Ye ſearching, ye *Newtonian* Angels! tell,
Where your great Maſters Orb? His Planets where?
Thoſe conſcious Satellites, thoſe Morning Stars,
Firſt-born of *Deity* from central Love.

Dr. YOUNG.

Many other Authorities might be produced from Writers of great Repute, were it neceſſary to trouble you with them †; but I believe thoſe above will be abundantly ſufficient for the preſent Purpoſe, if even an Apology were wanting for my own Conjectures. I ſhall therefore conclude this Letter with the following Paſſage out of *Pope*'s *univerſal Prayer*, and in my next ſhall proceed in the Work I have undertaken.

Yet not to Earth's contracted Span,
  Thy Goodneſs let me bound;
Or think thee Lord alone of Man,
  When thouſand Worlds are round.

I am, &c.

LETTER

† Particularly from *Fontenelle*, &c.

LETTER THE SECOND.

Concerning the Nature of Mathematical Certainty, and the various Degrees of Moral Probability proper for Conjecture.

SIR,

YOU know how much I am an Enemy to the taking of any thing for granted, merely becauſe a Perſon of reputed Judgment, has been heard to ſay, *it abſolutely is ſo*; an *Ipſe dixit*, and implicit Faith in ſome Caſes, may be both neceſſary and uſeful; but here, in Aſtronomy, I mean, every Man's Reaſon, by the Help of a very little Mathematicks, is able to bring wonderful Truths to Light without them; and Truths not only of the higheſt Importance to every Individual, but of a great and common Conſequence to all Mankind: And as ſuch, in all Ages of the World, have been judged worthy to be enquired into, by the beſt and wiſeſt of Philoſophers.

You are likewiſe very ſenſible how far the human Underſtanding is even at the beſt, from being infallible, and don't want to be told, how difficult it is in a Subject of this Nature to arrive at any tolerable Degree of Certainty, which before the Days of the ſagacious *Euclid*, and the penetrating *Archimedes*, was a Thing not to be expected. And many things which were then but barely Objects of Conjecture and Probability, have ſince been demonſtrated to be infallibly true. Time and Obſervation will undoubtedly, at laſt, diſcover every thing to us neceſſary to our Natures, and proper for us to know. As a Proof of which, we ſee human Wiſdom daily increaſes; and while a Capacity continues to make ourſelves ſtill more acquainted with the manifeſt Wiſdom and Power of GOD in the Works of his Creation, who is to tell us where to ſtop our Enquiries? Or who is ſo impious to ſet Bounds to a Science, which ſo evidently ſpreads through all Infinity, the Attributes of God, and an eternal Baſis for future Hope?

This Branch, or rather Body of Aſtronomy, I believe you will find to be quite new; and though evident Truths, are the principal Thing to be regarded in it, yet as being in its infant State, where lineal Demonſtration fails, as in ſome Caſes it cannot be otherwiſe, I hope you will give me Leave to make uſe of a weaker Way of Reaſoning, to convince you of the Point in Diſpute, I mean of that by the Analogy of known and natural Things.

I ſhall be extremely unwilling to affirm any thing for a *Fact*, or Truth, without bearing, if not the real Evidence, at leaſt a plauſible Reaſon, next to a Conviction, or moral Certainty, along with it; and therefore I will here endeavour to explain to you what I mean by moral Certainty and alſo by mathematical Proof.

Mathematical Proof, or Certainty, proper for Conjectures, may, to almoſt every Capacity, be illuſtrated as follows:

Suppoſe you had accidentally found a very ſmall Part of a viſibly broken Medallion, with nothing more expreſs upon it, than what is repreſented at *Fig.* 1. *Plate* I. a Perſon totally unacquainted with the mathematical Sciences, we may naturally conclude, would not be able to make any thing of it, or in the leaſt comprehend what it originally was, or meant; but if an Aſtronomer ſhould chance to ſee it, who of courſe we are to ſuppoſe knew the Order and Proportion of the planetary Orbits, he would immediately conclude, and with great Probability, on the Side of his Conjectures, that it might be Part of a Medal repreſenting the Solar Syſtem. In ſuch a Caſe may we not very naturally ſuppoſe he would reaſon thus?

The Arches A and B ſeem to be Portions of the reſpective Orbits of *Saturn* and *Jupiter*, and what may lead us to believe, that they are really ſo, and Part of the Solar Syſtem, is the oblique Curve C, which looks not unlike the Trajectory of a Comet.

This ſurely would be far from an irrational Conjecture, and conſequently in ſome Degree probable: But this is not ſufficient you'll ſay; To prove it we muſt have farther recourſe to the Mathematicks, and a Mathematician would immediately thus demonſtrate it to be true.

Firſt, by compleating the Circles geometrically from the fourth Book of *Euclid*, by the Aſſiſtance of any three Points E. F. G. the original Figure will be reſtored, as at *Fig.* 2. And ſecondly, by aſſuming any two Points, as F, E in the Curve C, if admitted a Parabola, by a well-known Problem in Conic Sections the Heliocentric Portion X. Y. Z. will eaſily be projected and ſhewn, as in *Fig.* 3. Laſtly, join this in Poſition to the former, and it will juſtly ſupply the Orbit, or Path of ſome one of the Comets; and if required, even what Comet may be diſcovered by comparing the Perihelion Diſtance Y. S. with their general Elements or Theories, in Dr. *Hally*'s *Synopſis* of the Motion of theſe Bodies. And if a farther Confirmation of the Truth of theſe Conjectures were wanting, the

PLATE. I.

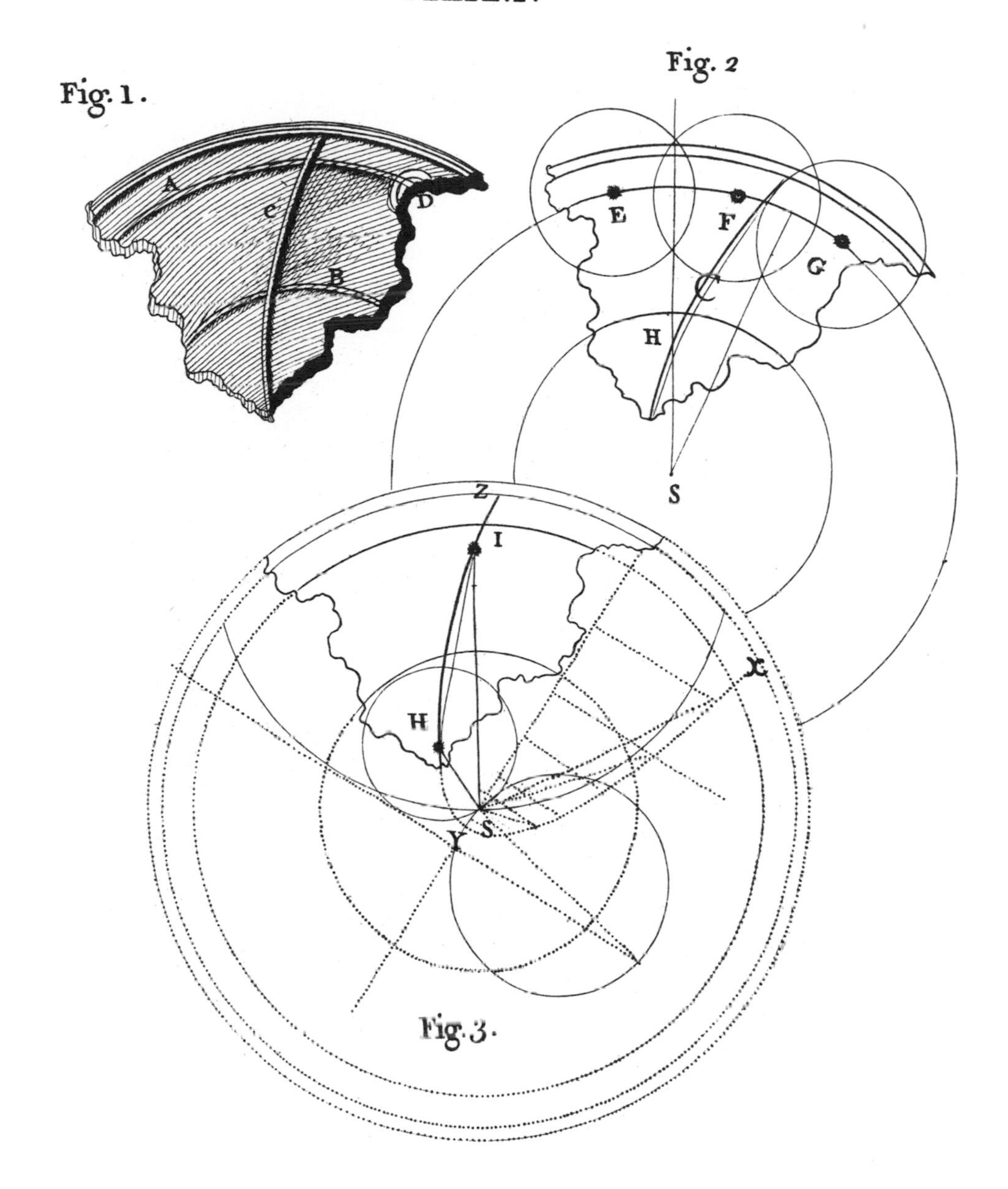

PLATE II.

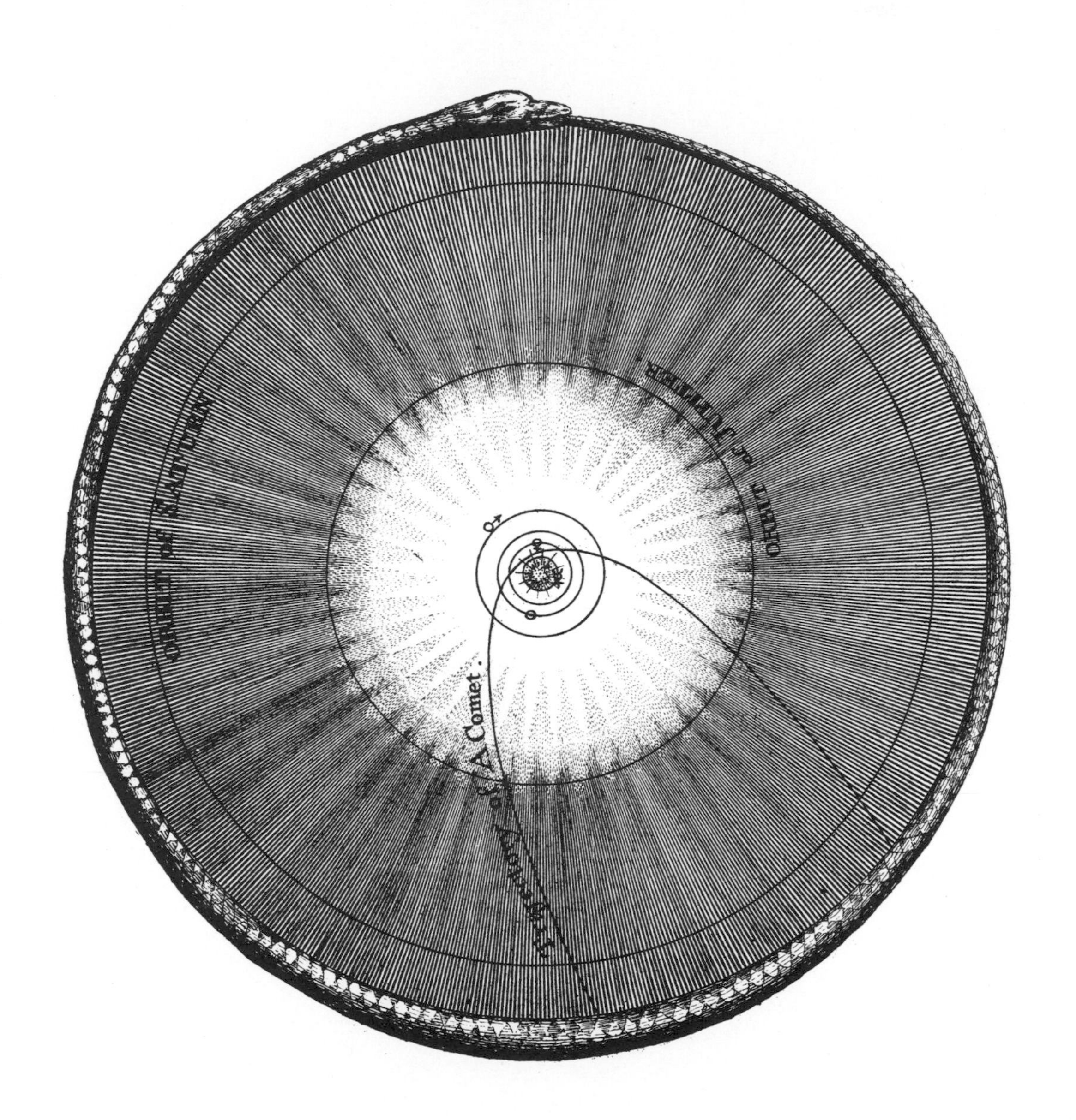

the ſmall concentric Circles at D would now be allowed beyond a Contradiction, to repreſent the ſecondary Orbits of *Saturn*; and thus the firſt Preſumption being carried thro' ſeveral corroborating Degrees of Probability, almoſt paſt a Diſpute, would become a mathematical Certainty; and the above imperfect Piece of Medallion, would evidently appear beyond a Contradiction to be Part of a Repreſentation of the ſaid ſolar Syſtem, and ſuch as is ſhewn in *Plate* II. Q. E. D. Thus in many Caſes, it often happens, that from a very ſmall Part of *orbicular Things*, we are able to determine the Form and Direction of the Whole: And hence you may conceive it no very difficult Taſk to a Mathematician, to deſcribe the Orbits of all the Planets in the ſolar Syſtem, though he had never obſerved them but in one and the ſame Sign of the *Zodiack*; thus far I have thought it would not be amiſs to explain to you the Nature of thoſe Steps, by which we arrive at moral Certainty, and where the Subject will admit of it, Mathematical Conviction, which will not a little contribute to ſtrengthen many of the Arguments hereafter made uſe of, and in ſome Degree ſerve to ſupply the Place of Proof, where infallible Demonſtration cannot from the Nature of the Thing be diſcovered.

But beſides the indiſputable Principles of Geometry, the univerſal Law of Analogy and Similitude of things, have a Privilege to aſſiſt us, in Conjectures relating to the heavenly Bodies, and though not of equal Force with the former, is often as concluſive as the Subject requires. This ſort of probable Evidence (as Dr. *Butler* obſerves,) is eſſentially diſtinguiſhed from " Demonſtrative by this, that it admits of Degrees; and " of all Variety of them, from the higheſt moral Certainty to the very " loweſt Preſumption; and that which chiefly conſtitutes Probability, is " expreſſed in the Word *Likely*, or Natural Likeneſs, as to State or Being." This general Way of arguing, I think, is allowed to be evidently natural, juſt and concluſive, and unqueſtionably to have its Weight in various Degrees, towards determining our Judgment: For Inſtance, ſhould any ignorant Perſon, endowed with rational Principles, cut open a *Pomegranate* of the natural Growth of *England*, and finding it full of ſmall Globules, or Kernels, upon being preſented with an every way ſimilar Fruit, ſaid to be the Produce of *Italy*, doubt of its being of the ſame Nature, and compoſed of like globular Seeds within; here indeed would be no mathematical Evidence to aſſiſt the Judgment, the Object of Proof being inviſible, but ſure from the external Similitude, the ſtrongeſt Probability of their being alſo internally the ſame. Again,

Is it natural to ſuppoſe, that the firſt Perſon who found a *Lark*'s Neſt, and in it ſeveral of the Female's Eggs, ſhould have any Apprehenſions of finding none in the *Nightingale*'s, only becauſe he had never ſeen one before, I believe the moſt illiterate Perſon of the earlieſt Ages, who had Curioſity enough for ſuch a Search, would be greatly diſappointed in ſuch a Caſe, and far from concluding that the *Nightingale* had none. Farther, ſhould any one who had ſeen ſeveral Sorts of Fiſh taken out of the River *Thames*, or out of the *Nyle*, have any ſort of Suſpicion that he ſhould find no ſuch Creatures in the *Seine* or the *Ganges*, though it ſhould be allowed that he had never ſeen any ſuch Creatures that were known to come from thence. Ocular Demonſtration, in ſuch a Caſe, would ſure be unneceſſary, and an Evidence of the firſt, I believe would be abundantly ſufficient to convince us of what we ought to look for at leaſt in the laſt: But then the Fiſhes of different Seas, and of Rivers are not of the ſame Species you'll ſay; but as it were infinitely diverſified through all the aqueous World, this is, and muſt be granted, and alike Variety of *Species* muſt alſo be granted, in the former Caſe of the Birds: But no Objection can poſſibly ariſe from any ſuch Diverſity, ſince we don't pretend to ſay, nor is it at all neceſſary, that the Beings in the ſidereal Planets ſhould be every where the ſame with theſe of our ſolar Syſtem, a Variety muſt every where be admitted, and will always be admired, where the Work is Nature's, and the Deſign GOD's.

All then that I here pretend to argue for, is a Univerſality of rational Creatures to people Infinity, or rather ſuch Parts of the Creation, as from the Analogy and Nature of Things, we judge to be habitable Seats for Beings, not unlike the mortal human.

Every Animal, and every Vegetable, that, as it were, naturally exiſts by the Virtues, Properties, or Laws of the mineral Kingdom, has ſomething of a ſecondary Nature, depending upon it as a Principle; and to ſay that the Stars, which are a certain viſible ſort of Cotemporaries in Space with the Sun, have no like planetary Bodies with ours moving round them, becauſe we cannot poſſibly ſee them, is no leſs abſurd and ridiculous, than to argue, that we can have no Reaſon to expect to find, in the proper Seaſon, Grapes upon every Vine — Figs upon every Tree — Roſes upon every Buſh — only becauſe ſome of them are at ſuch a Diſtance, that neither Roſe, Fig or Grape, can be diſcovered by the Eye.

This ſort of Reaſoning, though ſome perhaps may neglect it, I am perſwaded you will look upon as abundantly ſufficient for Things out of the Reach of Science to determine; and that the collective Body of Stars have not been diſcovered, to be together a proper Subject for ſuch Conjectures before, can ſurely only proceed from the Want of Time, neceſſary to compleat the Obſervations proper for a Foundation to build ſuch an Hypotheſis, or Theory upon. This is the great Article in which the Moderns have ſo much, and ever will have, an Advantage over the Antients. And hence it will appear, That

The

The Improvements and Discoveries of latter Ages are not at all owing to the greater Capacity of the Moderns, but from the Advantages received, or arising from the Inventions and Progress made by the Ancients. We at first in a manner walked by their Leading-strings, and though many of them now are broke, or useless, none can deny, but that formerly they were of great Advantage in promoting and directing philosophical Enquiries.

In an Assembly of the most eminent Men of all Ages, if we may suppose such a Conference amongst the illustrious Dead, on Purpose to deliver their several Sentiments familiarly together, on the most interesting Subjects of natural Knowledge, who would not lament the Disadvantages, poor old *Thales*, an *Hipparchus*, or a *Ptolomy*, would lie under, who had nothing but the Eye of Reason to direct them, in Opposition to the Judgment of a *Brahe*, or a *Galilæus*, who reaped so much Benefit from their compound Opticks? But on the other hand, perhaps if the solar System, was the Topic of Discourse, a * *Pythagorean* might very pertinently say to a *Newtonian*, "You have not gone much farther in the Light with our "Direction, than we did in the Dark alone; for you are still roving "round the same Circles." Much might be said upon this Head; but I believe it would be a difficult Matter to do Justice to all Parties: So here I intend to leave them, only must observe, that Posterity will always have the Advantage over their Predecessors; and that After-ages, in all Probability, will reap so great a Benefit from the Invention and Improvement of Fluxions, that scarce any thing, which is the immediate Object of such Enquiry, will long lie concealed from a true mathematical Genius.

For this, in which he has surpassed all the Antients, and greatly advanced the philosophical Sciences, the World is indebted to Sir *Isaac Newton*.

But as many of his Discoveries, such as relate particularly to the Laws of the planetary System, are but as so many Confirmations of the Conjectures and Imaginations of Astronomers and Philosophers before him, it perhaps will not be amiss to acquaint you a little with the Astronomy of the Antients concerning the Universe. And before I proceed to those of my own, shew you in the first Place how far their Speculations in the visible Creation have been carried; and with these I shall conclude this preparatory Epistle.

The Universe, or mundane Space, by which the Antients comprehend all Creation, has, from time to time, according to the Progress of Science, come under a sort of Necessity of being variously modell'd agreeable to the Opinion

* The true System of the Planets have been discovered above two thousand Years.

Opinion of the ſeveral Authors, who have judged themſelves wiſe enough to write upon it with a mathematical Foundation; And the coſmical Syſtem, by which is meant the Co-ordination of its conſtituent Parts has undergone almoſt as many Changes as its Elements are even capable of; every Age of the World, as Knowledge has increaſed, either from improved Imagination, or repeated Obſervations, producing ſomething new concerning it.

MILTON, no doubt, had all this Diverſity of Opinions in View, as appears from his ſuppoſed Pre-knowledge of *Raphael*, in the following Paſſage, *Book*. VIII.

> Hereafter, when they come to model Heaven,
> And calculate the Stars, how they will weild
> The mighty Frame! how build, unbuild, contrive
> To ſave Appearances, how gird the Sphere
> With centric and eccentric ſcribbl'd o'er;
> Cycle, and Epicycle, Orb in Orb.

8.79

But the following Synopſis, I believe, will abundantly convince you that from certain Obſervations only, we ought to form all our Notions of it, if we either hope to arrive at Truth, or expect our Ideas ſhould be ſupported by Reaſon.

ARISTOTLE was of Opinion, that the Univerſe, or Heaven, was all one World, and St. CHRYSOSTOM, TERTULLIAN, St. BONAVENTURE, TYCHO BRAHE, LONGOMONTANUS, KEPLER, BULIALDUS and TELLEZ, were of an united Opinion, that this one Heaven, or Univerſe, was all ſidereal and fluid. But AEGIDIUS, HURTADUS, CISALPINUS, and AVERSA, believing the ſame Heaven with them to be all one World, and that ſidereal, yet on the contrary held it to be ſolid.

CLEMENS, ACACIUS, THEODORET, ANASTASIUS, SYNAITA, PROCOPIUS, SUIDUS, S. BRUNO, and CLAUDIANUS MAMERTUS, ſuppoſed the univerſal mundane Space as divided into two Heavens, namely,

The Empyræum created the firſt Day,
And the Firmament created the ſecond Day.

Two Heavens were alſo held by JUSTIN MARTYR, the one ſidereal, and the other aerial. The firſt ſuppoſed by St. GREGORY NYSSENE, to be that of the fixed Stars, and the laſt, that of the Planets. But *Maſtrius* and *Bellutus*, though agreeing in the Number of Heavens, call one the *Primum Mobile*, and the other, the Starry Heaven.

Farther,

Farther, St. BASIL, St. AMBROSE, DAMASCENE, CASSIODORUS, GENEBRARDUS, SUAREZ, TANNERUS, HURTADUS, OVIEDUS, TELLEZ, and BORRUS, diſtinguiſhed the Univerſe as divided into three Portions, or Heavens.

	Or, as *Cajetan.*		*Tho. Aquinas.*	
The firſt called the Empyræum,		Watery,		
The ſecond ſuppoſed Sidereal,		Sidereal,		Watery,
And the laſt of all, Aerial.		Aerial,		Sidereal.

Again, St. *Athanaſius* adds to thoſe of the fix'd Stars, the Planets, and the Air, that of the *Empyræum*, and makes in all four Heavens.

But as the Number of the Heavens thus increaſes, and will become ſubdivided in the ſubſequent Account of them, to give you a better Idea of the Order of theſe celeſtial Portions of the mundane Space, it will not be amiſs to form what remains of them into regular Sections of their proper Spheres and Syſtems.

See *Plate* III. in which Figure, the firſt repreſents a Section of the coſmical Theory of *Oviedus* and *Ricciolus:* Both conſiſting of five Heavens, *viz.*

By *Oviedus*, ſidereal and ſolid.			By *Ricciolus*, ſider. and fluid.		
	The fixed Stars,	A		*Empyræum*, - -	G
	Saturn, - - - - - - -	B		The Water, - -	F
	Jupiter, - - - - - -	C		The fixed Stars,	A
	Sol, with ♂, ☿ and ♀ included	D		The Planets, -	H
	The Moon. - - - -	E		The Air. - -	I

Fig. II. repreſents that of venerable *Bede* and *Rabanus*, *viz.* of Seven Heavens.

And according to *Bede* compoſed of		But by *Rabanus*,
The Air, - - — - -	P	The Atmoſphere,
The Æther - - - - - -	O	The upper Air,
Olympus, - - - - - - -	N	The inferior Fire,
The Element of Fire, - - -	M	The ſuperior Fire,
The Firmament, - - - -	A	Sphere of the fixed Stars,
The Angelical Region, - -	L	The Chryſtalline Heaven,
Realm of the Trinity. - -	K	The *Empyræum.*

Fig.

Fig. III. Represents the Hypotheses of *Eudoxus, Plato, Calippus, Cicero, Riccius, Philo, Remigius, Aben-Ezra, Carthusianus, Lyranus, Tostatus, Brugensis, Orontius, Cremoninus, Philalethæus, Amicus,* and *Ruvius*; also the *Babylonians* and *Egyptians.*

Consisting of Eight Heavens,

All Sidereal, *viz.* The Sphere of the fix'd Stars, and those of the Seven Planets.

Fig. IV. is that of *Macrobius, Haly Alpetragius, Rabbi-Josue, Rabbi Moyses, Scotus, Abraham Zagutus, Sacroboscus, Claromontius, Avigra,* and *Arraiga.*

All of Nine Heavens,

Comprehend a *Primum Mobile* Q, or, according to *Arriaga,* a solid *Empyræum.* The Sphere, of fixed Stars A, and the seven Regions of the solar Planets.

Fig. V. is that of the great *Alphonsus, Fernelius, Regiomontanus, Amicus, Maurolycus* and *Langius*; also of *Azahel, Thebit,* and *Isaac Israelita*; and likewise of *Gulielmus Parisiensis,* and *Johannes Antonius Delphinus.*

Consisting of Ten Heavens, made up of

A *Primum Mobile* — — — S *Empyræum.*
A Sphere of *Tripidation* in Longitude — R *Primum Mobile.*
The Sphere of the fixed Stars — — A
And those of the seven solar Planets within.

Note, Some Authors place the Sphere of *Tripidation* in Longitude below that of the *Aplain,* or Eighth Sphere.

Lastly, Fig. VI. is the Heaven of *Petrus Alliacensis,* the College of *Conimbra, Martinensis,* (and sometime) of *Clavius*; and also *Johannes Warnerus, Leopoldus de Austriâ, Johannes Antonius Maginus*; and lastly, of *Clavius.*

In all Eleven Heavens containing,

T A *Primum Mobile,* or, as others say, an *Empyræum.*
V A Sphere of Libration in Latitude.
W A Sphere of Libration in Longitude.
A The Sphere of the fixed Stars, and those of the Planets.

Thus you see how many various Opinions have from time to time been imbraced concerning the Fabric and Formation of the visible Universe; all of which are now and have long been exploded; and although at first advanced by Men of the greatest Learning, and of the deepest Penetration in natural Knowledge, it does not appear from any one of their Opinions, that they had any the least Notion of infinite Space, but as it were

PLATE III.

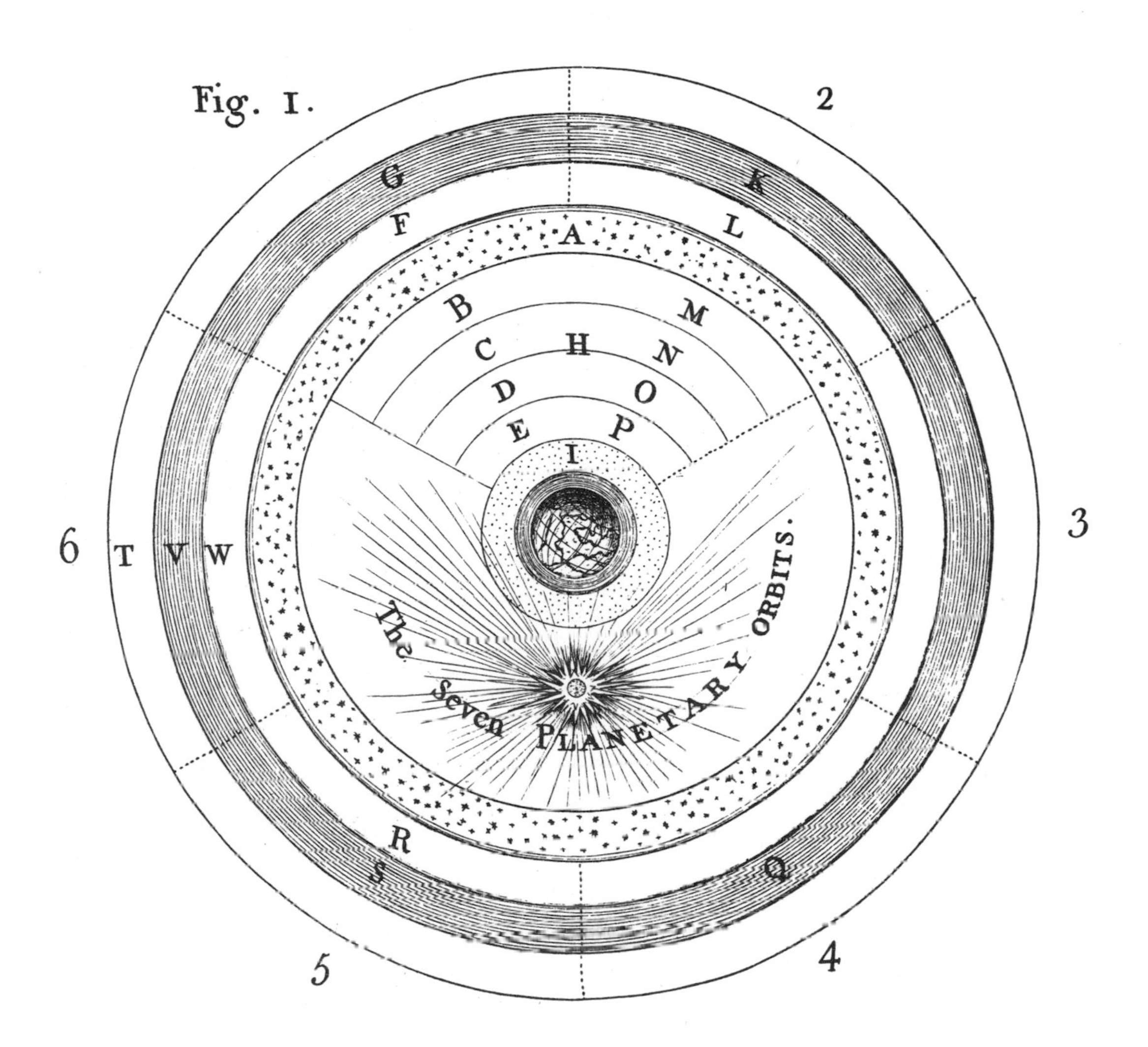

Fig. I.
2
3
4
5
6
G
K
F
A
L
B
M
C
H
N
D
O
E
P
I
T V W
The Seven PLANETARY ORBITS.
R
S
Q

were confined the Divine BEING to their limited Notions, as one may ſay in an Egg-ſhell. If therefore what I ſhall hereafter advance, extend ſo far without the known Creation, that you can poſſibly conceive no Bounds to the Works of infinite Wiſdom and Power, I hope you will be in no Danger of looking upon it as more ridiculous, or abſurd, than what ſo many of the wiſeſt Men of every Age have thought proper to attempt, and have judged worthy of their Attention ſo long before me. If any thing leſs ſo, I ſhall think myſelf happy enough in having broke, or rather paſſed the narrow Limits to which the Creation has for ſo many Years been confined, in hopes of tempting Men of greater Talents to look up wards, and purſue ſo noble a Subject as far as the human Underſtanding is capable of comprehending it.

To the Opinions above might be added many more, particularly that of *Johannes Baptiſta Turrianus*, and *Fracaſtorius*, who increaſed the Number of Heavens to fourteen, *viz.* ſeven on each Side the *Aplané*.

But of this I have ſaid enough; in my next I ſhall proceed to Matter better grounded,

And am, &c.

LETTER THE THIRD.

Concerning the Nature, Magnitude, and Motion of the Planetary Bodies round the Sun, &c.

SIR,

THE younger *Pliny*, if I remember right, ſomewhere ſays, that there is, or ought to be, a wide Difference betwixt writing to a Friend, and writing to the Publick: I have indeed pleaſed myſelf with the one, but am far from thinking myſelf qualified for the other; I muſt therefore rather intreat you, though perhaps you cannot poſſibly overlook all my Faults as an Author, to excuſe them at leaſt in the Friend, and by ſuch kind of unlimited Indulgence, you will give me a much greater Chance to do the Subject ſome Juſtice, though I own I deſpair in this firſt Attempt, to reconcile every thing I advance to your more cool and impartial Reaſoning. But to the Buſineſs:

As I have no Ambition to have the Subſtance of my Theory more admired by you than underſtood, which is too often the Caſe in Works of this Nature, I muſt beg leave to repeat to you Part of a former Diſcourſe, which will refreſh in your Ideas the principal Laws of the Syſtem of our Sun, and make you properly acquainted with ſuch Things as are neceſſary to be known in the now-eſtabliſhed Aſtronomy of * *Copernicus*, &c. before I proceed to any new Matter.

The

* NICOLAUS COPERNICUS, ſtiled by *Bulialdus*, *Vir abſolutæ ſubtilitatis*, was a Native of *Thorn* in *Poliſh Pruſſia*, and Canon of the Church of *Frawenburgh*; he was Scholar to *Dominicus Maria* of *Ferrara*, to whom he was Aſſiſtant in his aſtronomical Obſervations at *Bologne*, and Profeſſor of the Mathematicks at *Rome*, in his noble Work, *De Revolutionibus Orbium Cæleſtium*; he fortunately revived, happily united, and formed into an Hypotheſis of his own, the ſeveral Opinions of *Philolaus*, *Heraclides Ponticus*, and *Ecphantus Pythagoreus*, *viz.* after the Opinion of *Philolaus* he made the Earth to move about the Sun, as the Center of its annual Motion; and according to *Heraclides* and *Ecphantus*, he likewiſe gave it a diurnal Rotation round its own Axis: Which Syſtem has withſtood all Oppoſition; and as *Ricciolus*, (though a Diſſenter from it) obſerves, *Per damna, per cædes, ab ipſo ſumit opes, animumque ferro.*

The Sun, you are not to learn, is the reputed Center of our *Planetary Syſtem*, and may remember, that the Earth on which we live, and theſe five following *Erratic Stars*, viz. SATURN, JUPITER, MARS, VENUS and MERCURY, have been demonſtrated to move round him in the Order and Manner following.

Saturn is found to complete one Revolution round the Sun in twenty-nine Years, one hundred and ſeventy-four Days, ſix Hours, and thirty-ſix Minutes; at the Diſtance of about ſeven hundred and ſeventy-ſeven Million of Miles. *Jupiter* performs a like Revolution in about eleven Years, three hundred and ſeventeen Days, twelve Hours, and twenty Minutes; diſtant from the Sun about four hundred and twenty-four Millions of Miles. *Mars* compleats his Circuit in one Year, three hundred and twenty-one Days, twenty-three Hours, and twenty-ſeven Minutes; and his mean Diſtance is about one hundred and twenty-three Millions of Miles.

Theſe three are called ſuperior Planets, as being farther from the Sun than the Earth, and circumſcribing its Orbit.

The Earth circumambiates her Orbit in one ſolar Year, *viz.* in three hundred and ſixty-five Days, five Hours, forty-eight Minutes, and fifty-ſeven Seconds; at the mean Diſtance of eighty-one Million of Miles.

The Radius of *Venus*'s Orbit is about fifty-nine Millions of Miles; and that of *Mercury* nearly thirty-two Millions, *ditto.*

The Heliocentric Revolution of *Venus*, is made in two hundred and twenty-four Days, ſixteen Hours, forty-nine Minutes, and twenty-ſeven Seconds; and that of *Mercury*, in eighty-ſeven Days, twenty-three Hours, fifteen Minutes, and fifty-four Seconds. Theſe two laſt Planets are called inferior Ones, as being circumſcribed by the Earth.

The Diameter of the Sun being demonſtrated to be nearly ſeven hundred and ſixty-three thouſand Miles:

The proportional Magnitudes of all the above Planets will be found nearly as follows, *viz.*

	The Diameter of the Globe,	
Of *Mercury*	- - 4,240	Miles.
Venus	- - 7,900	
the Earth	- - 7,970	
Mars	- - 4,440	
Jupiter	- - 81,000	
and *Saturn*	- - 61,000	

D 2 Thus

Thus much I have thought proper to premiſe, and for your immediate Inſpection, have added the following Schemes, that nothing may be wanting to give a general Idea of the Order of the celeſtial Bodies in our own Syſtem, before I attempt to lead you through the neighbouring Regions of the Stars to the more remote Tracts of Infinity.

PLATE IV.

Is a true Delineation of the ſolar Syſtem, with the Trajectories of three of the principal Comets, whoſe Periods and Orbits have been accurately determined, and are repreſented in their true Proportion and Poſition to one another, and the Order of the Planets round the Sun, marked with their reſpective Characters, *viz.* ♄, for *Saturn*, ♃, *Jupiter*, ♂, *Mars*, ⊕, the Earth, ♀, *Venus*, and ☿, *Mercury*. The Scale being nearly five hundred and eighteen Millions of Miles to an Inch.

PLATE V.

Is a true Projection of the Syſtem of the known Comets; in which are repreſented nine of the chief Trajectories, from their *Aphelii* to their *Perihelii*, all in juſt Proportion and Poſition to the Orbits of *Saturn* and *Jupiter*, which are alſo repreſented by the two concentric Circles, ſuppoſed to be drawn round the Sun as their Center.

The Ellipſis, or Trajectory, marked A, ſhews the Poſition and Path of the Comet which appeared in the Year 1684, whoſe Period is ſuppoſed to be about fifty Years, and has been obſerved within the Region of the Planets once.

That mark'd B, is the Way of the Comet of 1682;
The Period conjectured to be about ſeventy-five Years and a half, and has been obſerved thrice.

C, Way of the Comet of 1337;
The Period about 100 Years, obſerved once.

D, That of the Comet of 1661;
The Period about 129 Years, obſerved twice.

E, Tract of the Comet of 1618;
The Period about 160 Years, obſerved once.

F, Way of the Comet of 1677;
The Period about 200 Years, obſerved once.

G, Way of the Comet of 1744;
The Period about 300 Years, obſerved once.

H, Way of the Comet of 1665;
The Period about 400 Years, obſerved once.

I, Way of the Comet of 1680;
The Period about 575 Years, obſerved thrice.

The

PLATE IV.

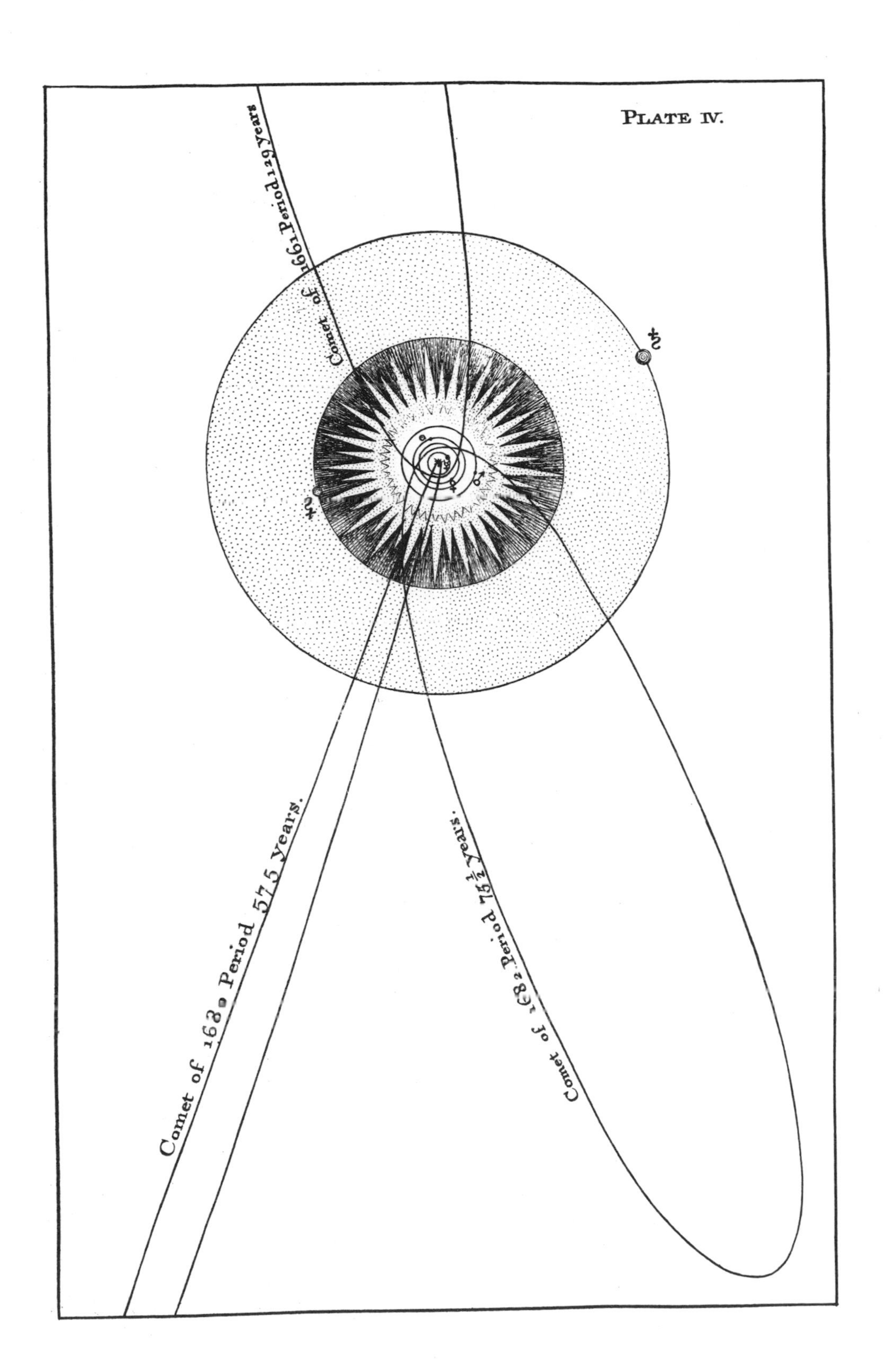

Comet of 1661. Period 129 years
Comet of 1680. Period 575 years.
Comet of 1682. Period 75½ years.

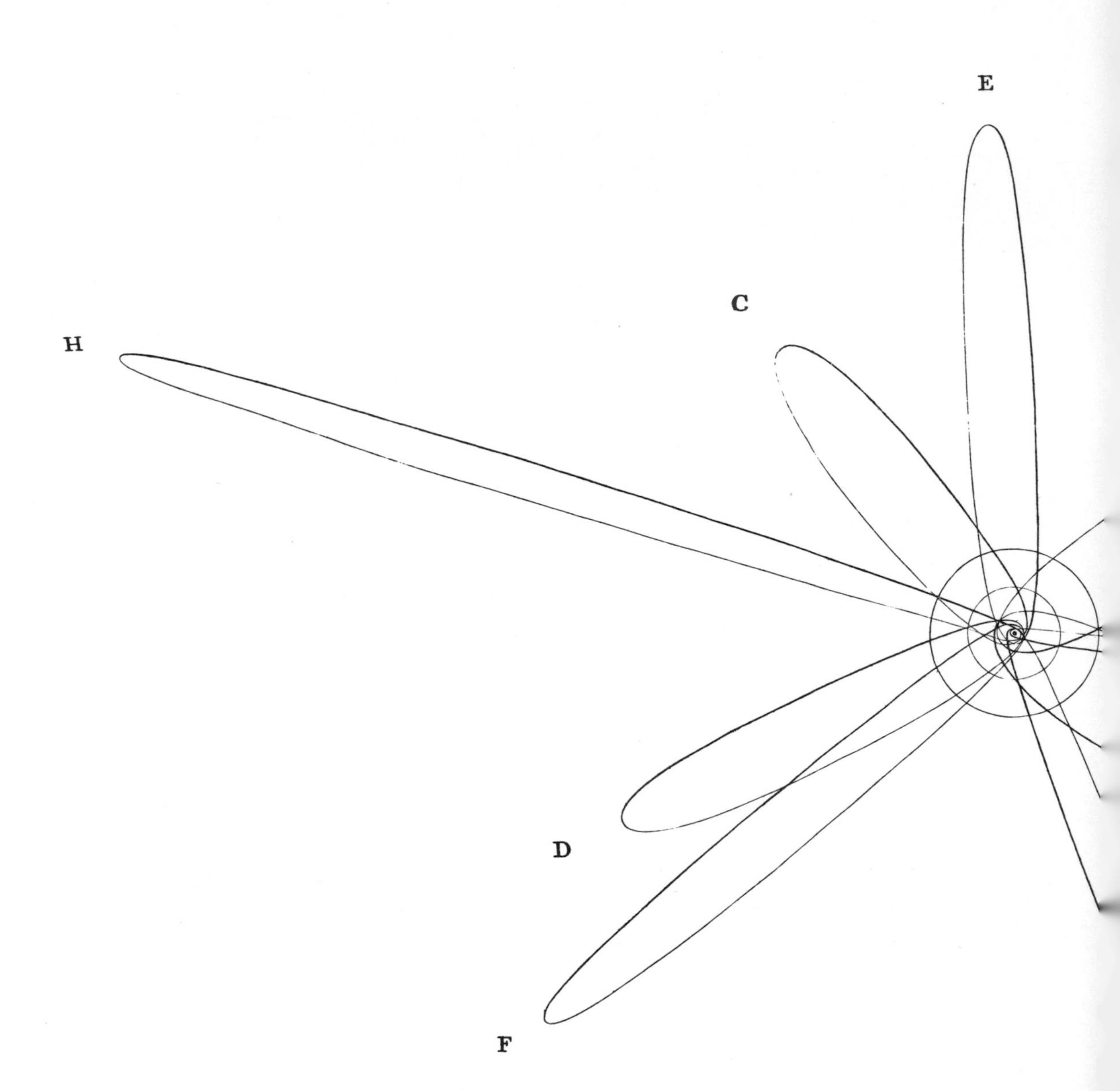

PLATE V.

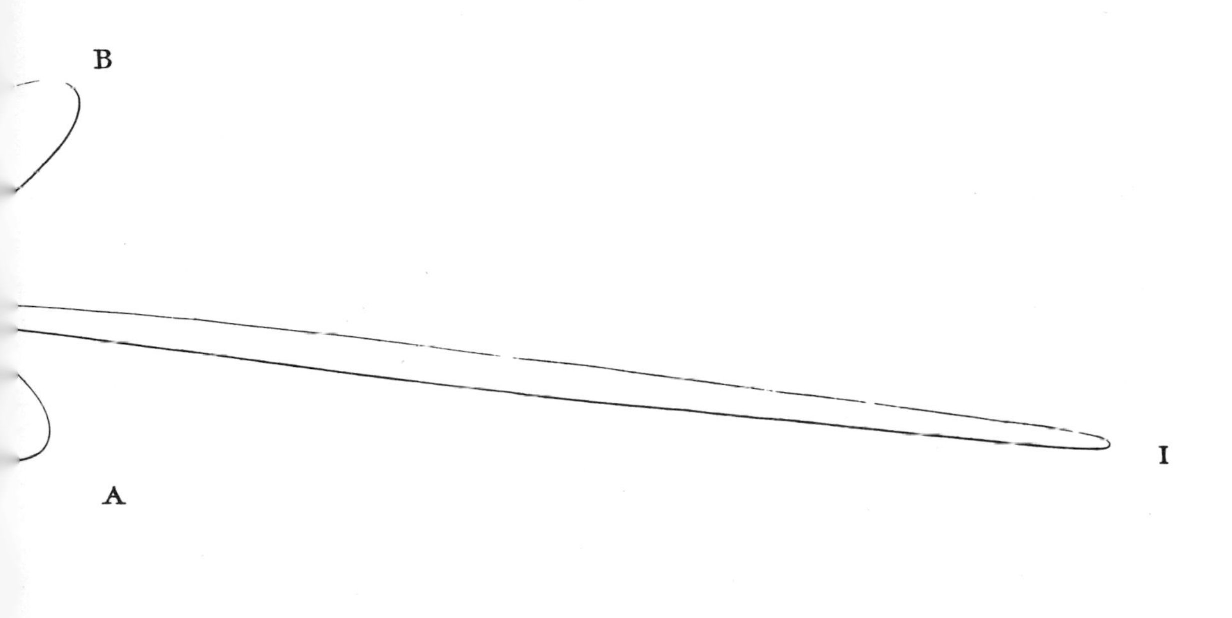

G

The Scale of this Syſtem is equal to one Third of the former.

Here I muſt obſerve to you, as a Thing I judge may prove of great Conſequence with regard to the Syſtem of Comets, which is as yet very imperfect: That I am ſtrongly of Opinion, that the Comets in general, through all their reſpective Orbits, deſcribe one common Area, that is to ſay, all their Orbits with regard to the Magnitude of their proper Planes, are mathematically equal to one another; which, if it once could be proved, and confirmed by Obſervation, the Theories of all the Comets that have been juſtly obſerved, might eaſily be perfected, and their Periods at once determined, which now we can only gueſs at, or may wait whole Ages for more Certainty of. What leads me to believe, that this may prove to be really the Caſe is this.

I find by Calculation, that the Orbits of the two laſt Comets, whoſe Elements have been moſt corrected by Sir *Iſaac Newton* and Dr. *Hally*, are to one another, according to their Numbers, nearly as * 13 to † 17, notwithſtanding one of them is one of the moſt erratick that ever came under our Obſervation; and the other one of the moſt neighbouring to the Sun.

But it is well known to all Mathematicians, that the firſt of theſe Comets moved in ſo eccentric a Trajectory, that the leaſt Error in its almoſt incredible Proximity to the Sun will produce a very ſenſible Difference in the Area of the Orbit: And accordingly, if we moderate the Perihelion Diſtance of this Comet, by making it but 1000 inſtead of ‡ 612, which is but increaſing it a $\frac{1}{35000}$th Part of the great Radius of the Orbit, (which is an Error every Aſtronomer will readily grant is very eaſily made) and we ſhall find the Orbits of the ſaid two Comets to be exactly equal.

Further, I muſt inform you, that the Comet of 1682, which the above compared with, ſeems to have been ſo accurately obſerved, that it does not appear to have altered its Perihelion Diſtance half a 68th Part in one intire Revolution. Now, if we can with any Show of Reaſon, and a Probability on our Side, bring the Areas of theſe two extream Comets, as I may call them, to an *Equality*, ſure we may conclude, it is a Subject highly worthy to be more conſidered and enquired into.

PLATE

* 1316539,968282 Comet of 1680.
† 1708155,4644 Comet of 1682.
‡ The Number in Dr. *Hally*'s Synopſis.

PLATE VI.

Is a true Repreſentation of the ſatellite Syſtems, proportionable to one another, and to the Orb of the Sun's Body, that a juſt Idea of the Diſtances of thoſe ſecondary Planets, may be eaſier had from their reſpective primary ones.

S repreſents the ſolar Body with its Atmoſphere. *Fig.* 1. is the Syſtem of *Saturn* from the ſame Scale. *Fig.* 2. that of *Jupiter* from *ditto.* And *Fig.* 3. the Orbit of the Moon round the Earth, in the ſame Proportion.

But as you can have but a very imperfect Idea of the Magnitude of theſe laſt Circles, with regard to the Body of the Earth or Moon,

PLATE VII.

Is a true Projection of their real Globes, at their proper Diſtance from each other, with their common Center of Gravity, and the Point and Line of equal Suſpenſion betwixt them, *viz.*

A, repreſents the Globe of the Earth.
B, that of the Moon.
C, Point, and C D, Line of equal Suſpenſion betwixt them.
E, Common Center of Gravity, which deſcribes the *Orbis Magnus.*
E, F, and B, G, is the Orbit of the Moon.

Farther, that nothing may be wanting to give a true Notion of the whole together,

PLATE VIII.

Is a proportional Drawing of all the primary and ſecondary Planets together, diſtinguiſhed by their Characters, proper to attend a Globe of twelve Inches Diameter, ſuch a one being ſuppoſed to repreſent the Sun.

PLATE IX.

Is an exact Scheme of the principal known Comets, in juſt Proportion, to the Globe of the Earth repreſented at A, with the Nuclus, and Part of the Tail of the Comet of 1680, B, as it was obſerved in its Aſſent from the Sun, *viz. a a* the Comet's natural Atmoſphere, *z z z*, the *Denſer Matter* winding itſelf into the Axis of the Train *x x*, the inflam'd Atmoſphere and Tail dilated near the Sun. C, repreſents the Ball of the Comet of 1682, D, that of 1665, E, that of 1742, and F, the Head of the Comet of 1744.

And again, that you may have ſome Notion of the apparent Magnitudes of all theſe Planets and Comets, &c. as they appear at the Earth,

PLATE

PLATE. VI.

Figure. I.

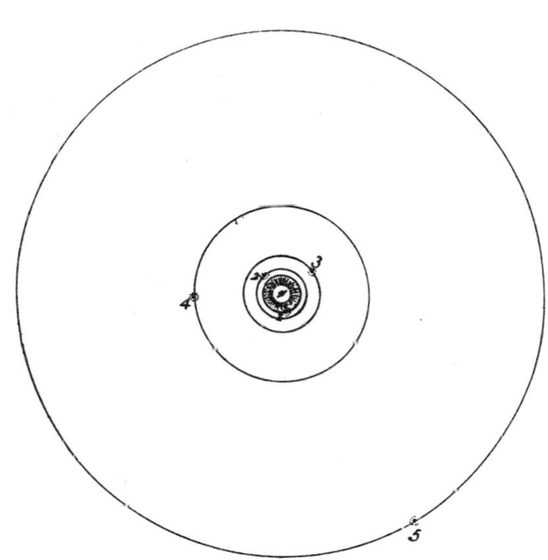

S

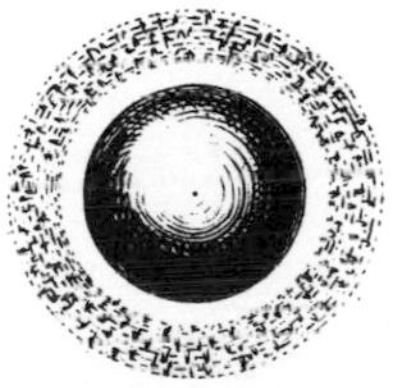

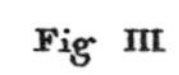

Fig III

Fig. II.

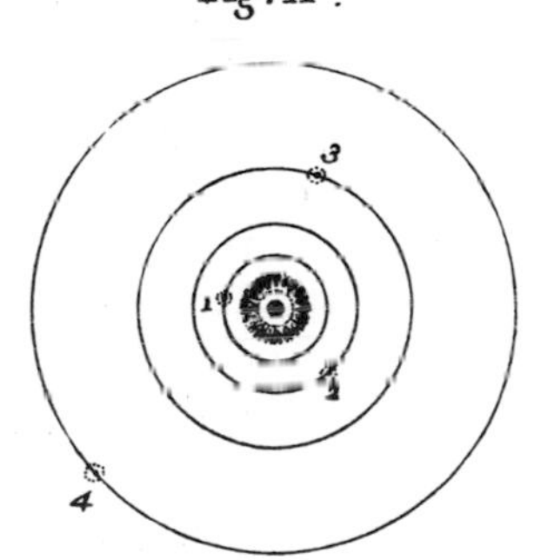

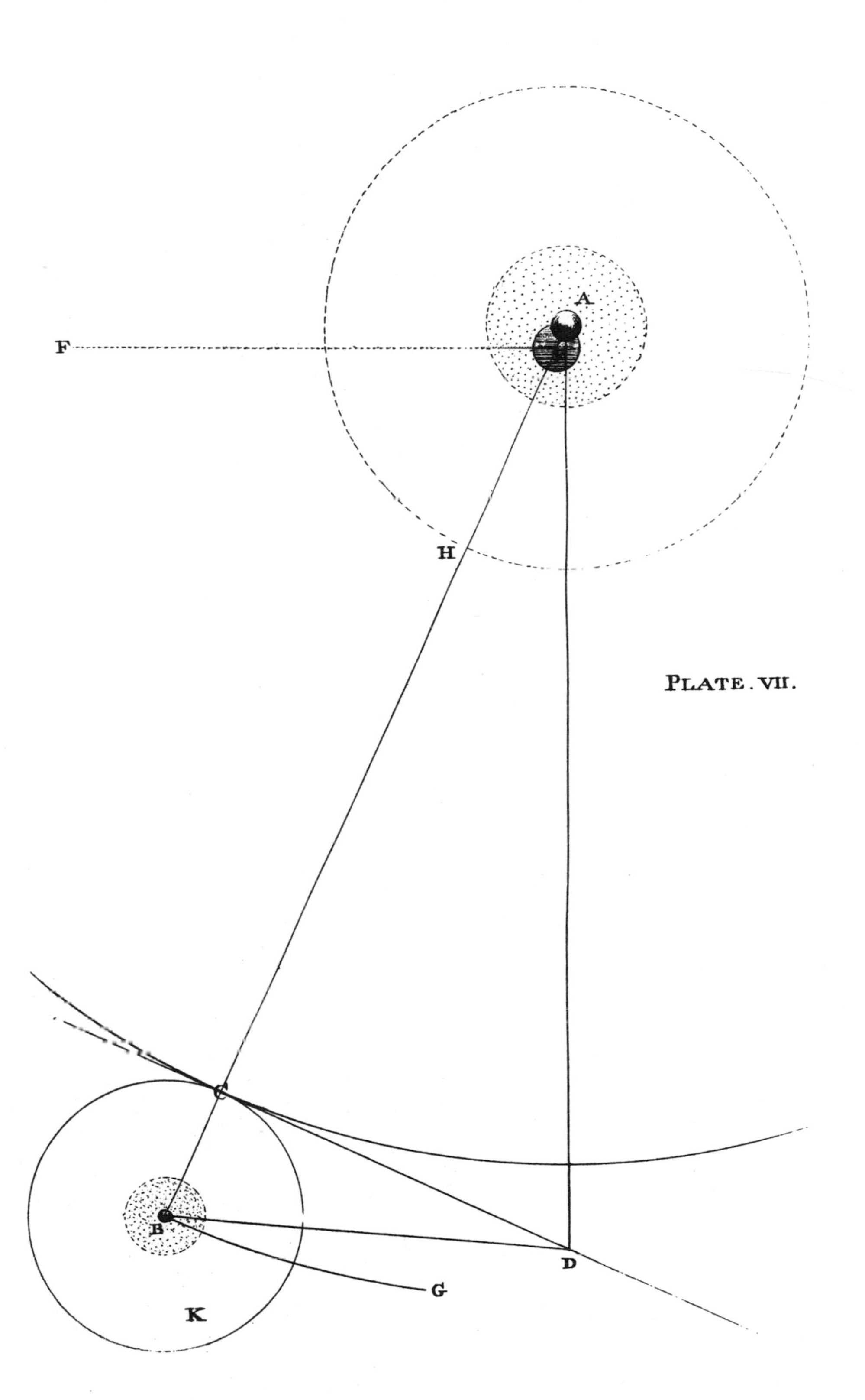

PLATE. VII.

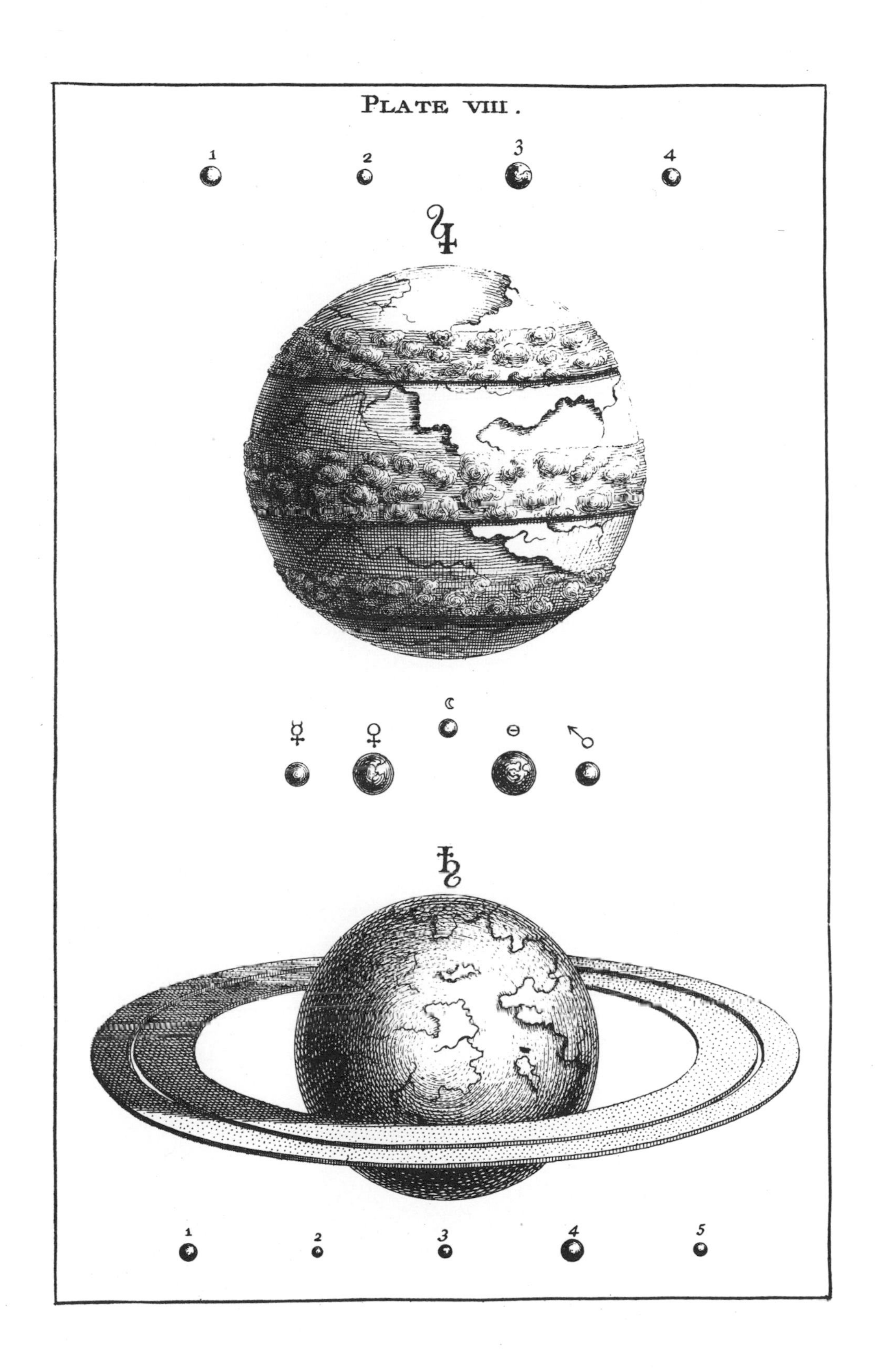
PLATE VIII.
1
2
3
4
♃
☿
♀
☾
⊖
♂
♄
1
2
3
4
5

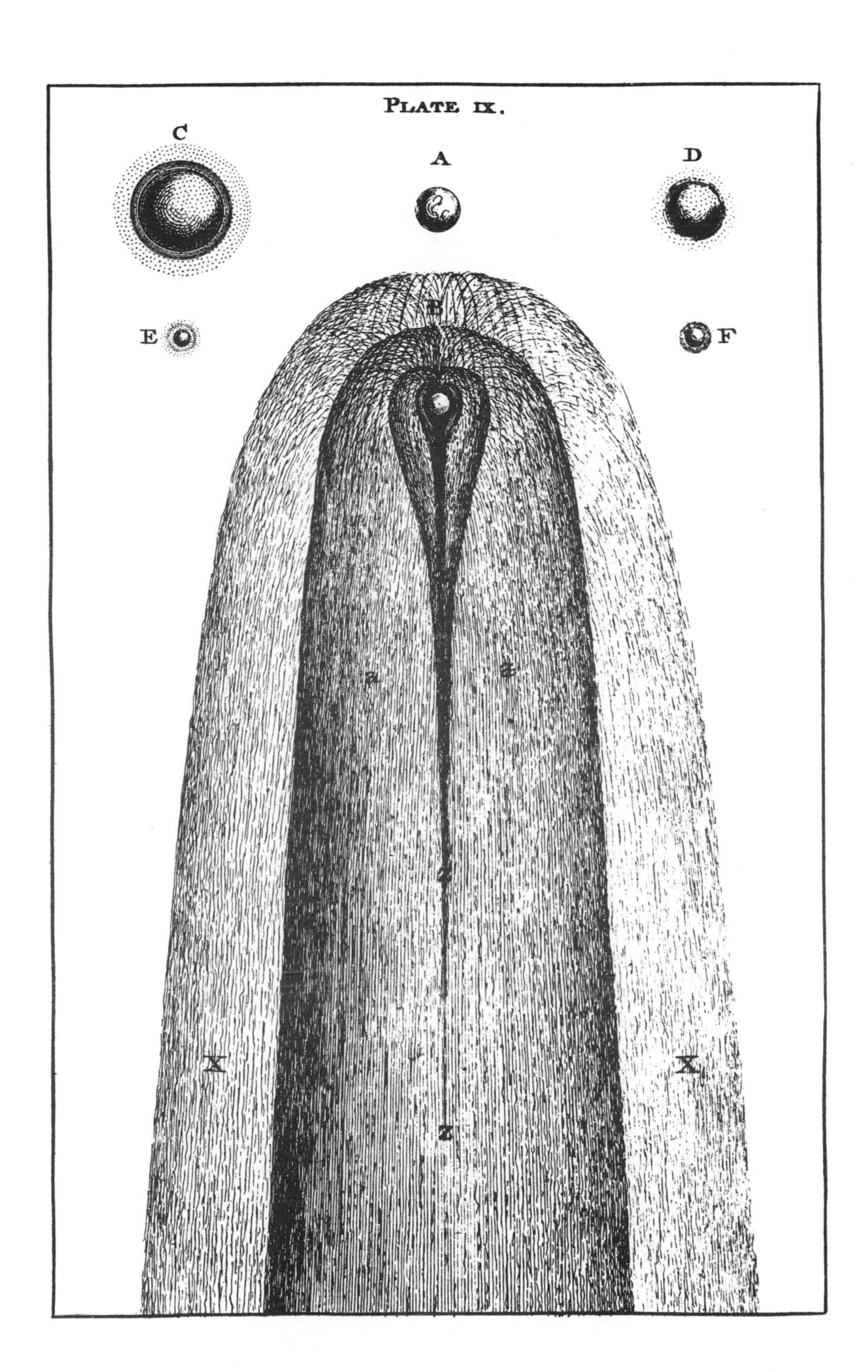

Plate IX.

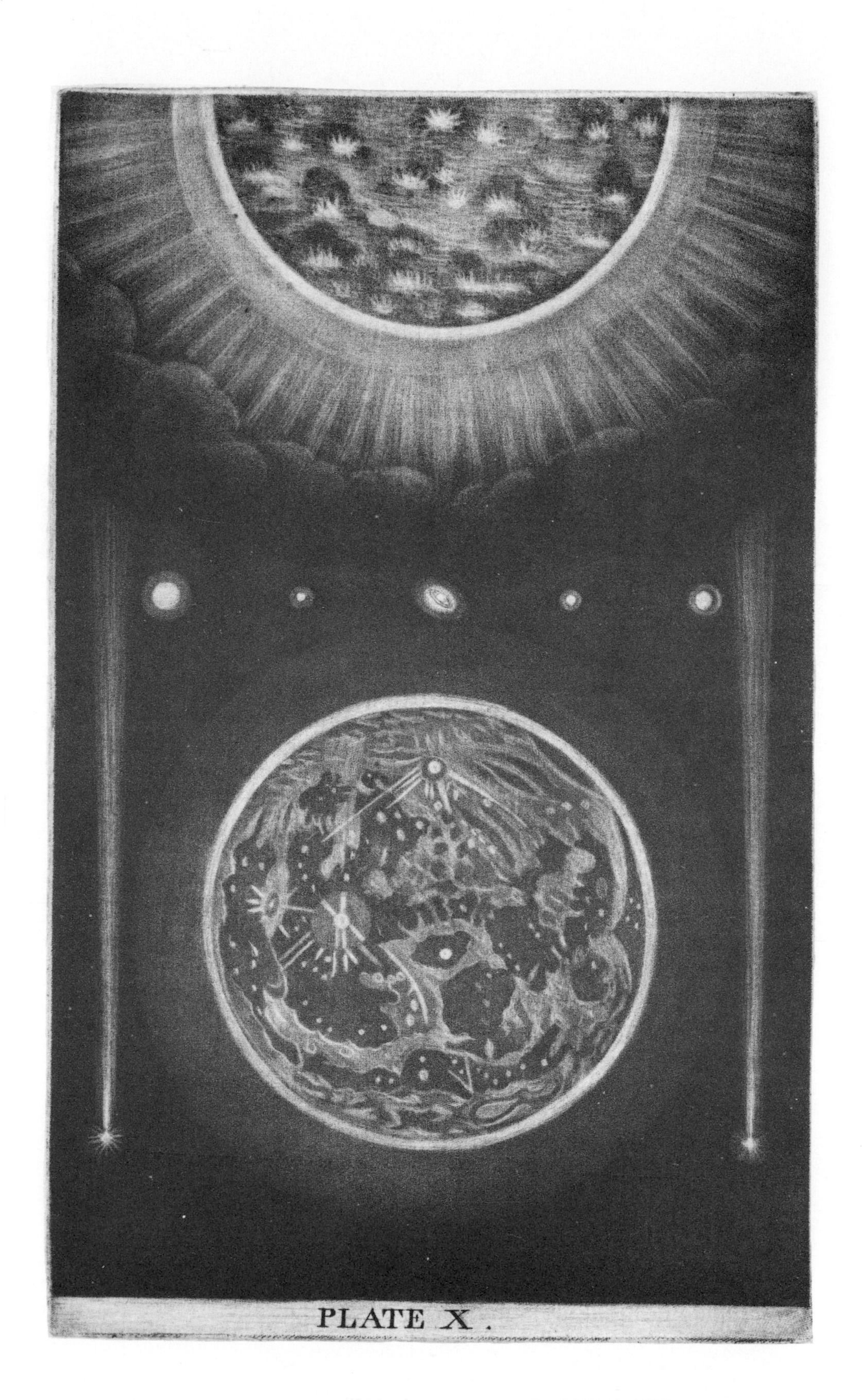

PLATE X.

PLATE X.

Repreſents the Sun and Moon in the juſt Proportion of their mean Diameters, with two of the Comets A and B, and the five erratick Planets, as they are obſerved at the Earth, in a middle State of their Diſtances from it.

For a more full and particular Deſcription of all the Parts of the ſolar Syſtem, and of the home Elements of Aſtronomy in general, I refer you to my *Clavis Cæleſtis*, &c. where every thing concerning the Planets, Comets, and Stars; and their real and apparent Motions, are at large repreſented, explained, and accounted for, for the Benefit of ſuch as have not made the Mathematicks their regular Study.

Now, to convince you that the Planets are all in their own Nature no other than dark opaque Bodies, reflecting only the borrowed Light of the Sun, I muſt recommend to your Obſervation, this natural and ſimple Experiment, which almoſt any Opportunity of ſeeing the *Moon* a little before the Full, will put into your Power to make; but beſt and eaſieſt when the Sun is in any of the North Signs, *i. e.* in *Summer*.

At ſuch a time, the Sun being near ſetting, the Moon will appear in the eaſtern Hemiſphere; and if there be any bright Clouds northward, or ſouthward near her, you will plainly perceive, that the *Light* of the one is of the ſame Nature with that of the other; I mean the Light of the Moon, and that of the Cloud. To me there never appeared any Difference at all; and I am perſwaded, were you to make but two or three Obſervations of this kind, which is from Nature itſelf, a ſort of ocular Demonſtration, you cannot fail of being convinced, that the Moon's Light, ſuch as it is, without Heat, can poſſibly proceed from no other Cauſe than that which illumines the Cloud: For if the Clouds, whoſe Compoſition we know to be but a thin light Fluid, formed of condenſed Vapours only, is capable of remitting ſo great a Luſtre, how much more may we not allow the Moon, which, Length of Time, and many other Circumſtances, have long confirmed to be a durable and ſolid Body.

The Increaſe of her Luſtre, indeed, during the Abſence of the Sun from us, to a leſs penetrating Genius than your's, may poſſibly afford ſome trifling Ground of Objection to the above Concluſions, as being drawn from the Phænomena of Day-light only; by reaſon in the Night, we have no Clouds in equal Circumſtances to compare with her.

But this I need not tell you, is all owing to her being ſeen through a darker Medium, and not to any real Increaſe of natural Light emitted from the Sun. As a Proof of which, were it neceſſary, you need only, ſhut out the Rays of the Atmoſphere, by the Help of a ſufficiently long

long Tube; and the Moon, or any other celeſtial Body, will appear through it, as bright in the Day-time as in the Night.

Thus all light Bodies of inferior Luſtre,. whether ſhining by their own natural Radiences, or by a borrowed Reflection, partake of the ſame Advantage, when removed from the more potent Influence of a ſuperior one; and hence it is, that the *Aura Ætherea ſhines out moſt manifeſt, when the Body of the Sun himſelf is hid, the Stars, and the *Via Lactea* moſt lively and numerous in the Abſence of the Moon, and thoſe Exhalations, or Meteors, vulgarly called Falling-ſtars, become only viſible (like Glow-worms) in the Night.

Here it may not be improper to tell you, that the Clouds are to us in effect no other than as ſo many Moons, whereby we have our artificial Day prolonged to us ſeveral Hours after the Sun is ſet, and likewiſe produced as much ſooner before he riſes; and were they to aſcend by ſtill ſtronger Power of Exhalation to an Elevation, all round the Atmoſphere, ſo as to form a Sphere equal to four Times the Globe of the Earth, there would then be no ſuch Thing as real nocturnal Darkneſs to any Part of the World.

The lunar Light then we may very juſtly conclude, proceeds originally from the Sun: And notwithſtanding many more Arguments might be drawn from the Demonſtration of her Phaſes, Eclipſes, &c. to prove it, yet none of them need here be added, to what has been already ſaid, to convince you of the Truth of it. This being granted, let us now conſider what Effect this, or a like Quantity of borrowed Light, would have, when removed to a much greater Diſtance.

I may, I think, ſuppoſe, that you know ſo much of Opticks as to underſtand, that all viſible Objects apparently decreaſe in Magnitude, as their Diſtance from the Eye increaſes. Conſequently, that, if the Moon's Orbit was placed as far again from the Earth as it really is, her Globe, or rather *Diſk*, would then ſeem to be but half as big as to us ſhe now appears to be, and of courſe ſtill farther, were ſhe placed at ten times the Diſtance ſhe is known to revolve at, her apparent Diameter would be reduced to a tenth Part only of what it now appears to be in her preſent Orbit, that is, one hundred Times leſs in viſible Magnitude than her neighbouring Diſk is found to be where it now is ſeen. And ſuch, but ſomething leſs, the two Planets *Venus* and *Jupiter*, which are frequently, in their Turns, our Morning and Evening Stars, appear to be through a common Teleſcope.

Now

* An *Helios*, or golden Light, always attending the Sun, and ſuppoſed to ſpread itſelf all round his Body in the Direction of his Equator, was very viſible during the total Darkneſs of the Eclipſe of 1715, and may be always ſeen about the Autumnal Equinox.

Now theſe two Planets, together with the other three, which we find moving in regular Orbits round the Sun, are all found ſubject to the ſame * Changes of *Phænomena*, in their various Aſpects with the Sun; and who can doubt but that they are all of the ſame or like Nature? But you'll ſay, perhaps, how are we ſure that *Venus* and *Jupiter* have no native Light of their own, ſince many of the ancient Philoſophers, and in particular *Anaximander*, allowed even the Moon to have ſome; and beſides, in Philoſophy, as well as in Logick, I think you hold there is no proving a Negative, at leaſt at ſuch a Diſtance.

To make you conceive the Impoſſibility of ſuch a Light, and next to a Demonſtration, convince you of the Unnaturalneſs of ſuch a Suppoſition, I muſt put you in mind, that ſome time ago, when I was laſt in the Country with you, I think it was about the latter End of Autumn, near the Winſter Solſtice, as we were walking one Evening, I bid you take notice of the Moon, which was then near ſetting, and about two Days old. You may remember, her whole Globe appeared to us very conſpicuouſly within a manifeſt Circle. You immediately told me, that that kind of Phænomenon the Country People called a *Stork*, or the old Moon in the new one's Arms. This I then endeavoured to explain to you, and I think made you ſenſible it was intirely an Effect of the Earth's, and an Appearance always to be expected at that Time of the Year. The Earth being then in the State of a Full-Moon to that Part of the lunar Orbit, and near her Perihelion, at which time, the Earth ſends back a Reflection to the † Moon twenty-five times more potent than that of the Moon to us.

Now the Planet *Venus*, from undeniable Principles of Geometry, is allowed to be nearly ſuch another Globe as the Earth is; and ſince the Earth, as I have juſt now related, is found to reflect much more Light to the Moon, by reaſon of her ſuperior Magnitude, than the Moon can poſſibly reverberate to Earth again; and ſince alſo 'tis plain, the Earth has no Light of its own, why then ſhould we imagine *Venus* to be endowed with a Luſtre, which we can prove to be no more than a ſimilar Body, and governed by the ſame Laws as the Earth is?

Anaximander's Miſtake, in ſuppoſing the Moon in ſome ſmall Degree a radiant Body of itſelf, lay, in not conſidering, that the faint Illumination here deſcribed, and viſible all over her Globe, ſoon after almoſt every Conjunction with the Sun; and probably in Eclipſes, alſo proceeded from the Earth; but the thing I think is too evident to expect any ſort of Contradiction,

E

* *Venus* and *Mercury* in every Heliocentrick Revolution, perform all the Changes of our Moon in a like Gradation and Defection of Light, both horned and gibos'd.

† Their Diameters being nearly as 1 to 5.

tradiction, therefore I hope you will admit it as a Truth, and consequently take it for granted, that the planetary Bodies in general, are meer terrestrial, if not terraqueous Bodies, such as this we live upon; which is the Thing I have chiefly in this Letter attempted to demonstrate, or have rather explained; and now I hope, for the future, you will receive the Idea of a Plurality of Worlds more favourably, and look upon astronomical Conjectures in a less ridiculous Light than you used to do, especially since you must allow, they give our unlimited Imaginations a like all endless Field of Contemplation, not only full of the wonderful Works of Nature, but also of a visible Providence.

I think I cannot conclude this Letter to you more properly, than with the following fine Lines of Mr. *Addison*'s from the *Spectator*, Vol. VI. No. 465. which I hope you are not so polite as to look upon as an unfashionable Quotation.

The spacious Firmament on High,
With all the blue ethereal Sky,
And spangl'd Heav'ns, a shining Frame,
Their great Original proclaim:
Th' unwearied Sun, from Day to Day,
Does his Creator's Pow'r display,
And publishes to ev'ry Land
The Work of an Almighty Hand.
Soon as the Ev'ning Shades prevail,
The Moon takes up the wond'rous Tale,
And nightly to the list'ning Earth,
Repeats the Story of her Birth:
Whilst all the Stars that round her burn,
And all the Planets in their Turn,
Confirm the Tidings as they roll,
And spread the Truth from Pole to Pole.
What though, in solemn Silence, all
Move round the Dark terrestrial Ball?
What tho' nor real Voice nor Sound
Amid their radiant Orbs be found?
In Reason's Ear, they all rejoice,
And utter forth a glorious Voice,
For ever singing, as they shine,
" *The Hand that made us is divine*."

And am, &c.

LETTER

LETTER THE FOURTH.

Of the Nature of the heavenly Bodies continued, with the Opinions of the Antients concerning the Sun and Stars.

S I R,

YOU tell me you begin to be a tolerable good *Copernican*, and would now be glad to have my Opinion further upon the Nature of the Sun and Stars, with regard to the Suggeſtion of their being like Bodies of Fire. This you ſay will go a great Way towards confirming you in the Notion you have begun to embrace of a Plurality of Syſtems, and a much greater Multiplicity of Worlds than our little ſolar Syſtem can admit of. Beſides, ſhewing in a very evident Light, that the Authorities cited in my firſt Letter are founded upon the cleareſt Reaſon.

Anaxagoras, you ſay, believed the Sun to be a Lump of red-hot Iron; *Euripides* thought it a Clod of Gold; and others ſtill more ridiculouſly have imagined it to be a dark Body, void of all Heat. That the Sun is a vaſt Body of blazing Matter, notwithſtanding the various Opinions of thoſe primitive Sages, will, I think, hardly admit of a Queſtion: Since the known Warmth of his prolifick Beams, and the viſible Effect of the Burning-glaſs, puts it quite out of the Power of our preſent Set of Senſes, at leaſt to argue againſt it; and how reaſonably we may imagine the Stars to be all of the ſame or like Nature, will ſufficiently appear from theſe following Conſiderations: Firſt, it is well known to all Mathematicians, that any viſible Object of any determined Magnitude may be reduced to the Appearance of * a phyſical Point, by removing the Eye of the Obſerver to a proper or proportionable Diſtance from it, within the finite View: And that the apparent Diameter of every luminous celeſtial Body, will always be diminiſhed reciprocally, in Proportion to the Diſtance from the Eye, till they become altogether imperceptible.

E 2 Thus

* What is here meant by a phyſical Point, is a Point viſible to the naked Eye, which human Art cannot divide; and ſo far it partakes of the Property of a mathematical one, which is only to be conceived, and not ſeen.

Thus the Diſk of the Sun, which appears to us at Earth under an Angle of about half a Degree, if ſeen from the Planet *Saturn*, would appear not much bigger than the Planet *Venus* or *Jupiter*, in their moſt neighbouring Vicinity does to us; and conſequently to an Eye placed in the Aphelion Point of the Orbit of the great *Comet* of 1680, his apparent Diameter would be ſo reduced as to ſeem but little bigger than the largeſt of the Stars; and by the ſame Analogy, or Way of Reaſoning, admitting Space and Diſtance infinite, which I humbly apprehend is not to be diſputed, were all the Matter in the Univerſe united, and conglobed in one Maſs, with reſpect to ocular Senſation, it might be diminiſhed ſo near to a mathematical Punctum, as to be almoſt adequate to our Ideas of Nothing.

This to any tolerable Optician, muſt be an evident Conviction of the Truth of the modern Aſtronomy, which now univerſally allow all thoſe radiant Bodies the Stars to be of the ſame Nature with the Sun; and that as certainly they are no other than vaſt Globes of blazing Matter, all undoubtedly ſhining by their own native Light.

But as you have often objected to what has been ſaid of the Diſtance of the Stars in general, and may poſſibly from a Suppoſition, that they are, or may be, much nearer to us, infer, that their Light, like that of the Planets, may be alſo borrowed from the Sun, or from ſome other radiant Body, which, from the Nature of the Suppoſition, muſt of Conſequence be inviſible to us, I judge it will not be amiſs to throw a few demonſtrative Arguments in your Way, in order to lead you a little out of the Path of an early Prejudice, and draw you as it were by Degrees through the Dawn of aſtronomical Reaſoning, out of your original Error, and reſcue your Imagination from the falſe Notions imbibed from Phænomena only in your younger Years. This I gueſs cannot fail of reconciling you to this more rational Way of Thinking, and make you acquainted with Truths of much Conſequence, which perhaps you have yet been an intire Stranger to. The grand *Deceptio Viſus*, which I muſt firſt endeavour to remove, and which as a ſort of Paradox in Nature, has, as I may ſay, impriſoned the Underſtanding of many ſuperficial Reaſoners, and in general all incurious Men, is this.

Moſt People are too apt to think originally, that as the Heavens appear to be a vaſt concave Hemiſphere, that the Stars muſt of courſe, as of Conſequence, be fixed there, like ſo many radiant Studs of Fire, of various Magnitudes; and take it for granted, chiefly deſigned for no other Purpoſe than to deck and adorn the Canopy of our Night. This was long ago the Opinion of *Thales* the *Mileſian*, and wants not the Authority of many

many of the Antients to back it. Others, in particular * *Ptolomy* of *Pelusium* in *Africa*, who from his Experience in this Science, is called by some the Prince of Astronomers, believed them to be Loop-holes in the vast solid celestial Firmament, emitting the Light of the Crystalline Heaven through it to all within it. The famous *Diogenes*, Cotemporary with *Plato*, conceived them to be of the Nature of Pumice-stones, and inclined to an Opinion, that they were the *Spiracula*, or Breathing-holes of Heaven. *Anaxagoras* thought them Stones snatched up from the Earth by the Rapidity of its Motion, and set on Fire in the upper Regions above the Moon.

But how ridiculous and absurd all these Opinions and Conjectures really are, will easily appear, if we but once consider the Nature of an unbounded Æther, and the amazing Property of infinite Space.

This, with what has been said before, will not a little assist your Imagination towards conceiving the Reasonableness of the Notion modern Astronomers are now confirmed in, of their being absolutely so many burning Balls, and which was no doubt, many Years ago, the Opinion of *Manilius*, as is evident from these Lines in his Poem of the Sphere.

For how can we the rising Stars conceive
A casual Production; or believe
Of the chang'd Heav'ns the oft renascent State
Sol's † frequent Births, and his quotidian Fate.

SHERBURNE.

And again in the same Poem:

The fiery Stars, and Æther that creates
Infinite Orbs, and others dissipates.

Zoroaster,

* *Ptolomy* supposed two Heavens above that of the fixed Stars, which he called the eighth; *viz.* a ninth, the Crystalline, and a tenth the *Primum Mobile*. See Letter the second.

The sacred Sun, above the Waters rais'd,
Thro' Heav'ns eternal, brazen Portals blaz'd;
And wide o'er Earth diffus'd his chearing Ray,
To Gods and Men to give the golden Day.

HOMER.

† *Xenophanes* believed the Stars to be no other than Clods set on Fire, quenched in the Day-time, and rekindled in the Night.

Zoroaſter, the firſt of all Philoſophers we read of who ſtudied the Stars, is reported to have believed them of a fiery Nature. *Empedocles* judged them to be Fire æthereal, ſtruck forth in its Secretion, and blazing in the upper Regions. *Plato* thought them Fire, with the Mixture of other Elements as Cements. *Heraclides* Worlds by themſelves, of *Earth*, *Air*, and *Fire*; and *Ariſtotle*, ſimple Bodies of the Subſtance of Heaven, but more condenſed.

But that I may not take up too much of your Time with Opinions that has been imbibed in the Infancy of Aſtronomy, and has long ago been exploded, I ſhall attempt but one Thing more to confirm your Sentiments in this new Doctrine.

Firſt, that the Stars are all at a Diſtance, not to be determined by the utmoſt Perfection of human Art, is manifeſt from their having very little, or no ſenſible Parallax; and conſequently, that any one of them is abſolutely bigger or leſs than another, from the ſimple Laws of Opticks, cannot poſſibly come under our Obſervation to be aſcertained; but that they all of them may be nearly of the ſame Size or Solidity, is as impoſſible, with any Shew of Reaſon to deny, ſince it is a known Principle in Geometry, that all viſible Objects naturally diminiſh, as has been ſaid before, or are magnified in a certain Proportion to their Diſtance from the Eye; and hence we may conclude, and not without Reaſon in its ſtrongeſt Light to ſupport us, that the ſmalleſt Stars, to the very leaſt Denomination, are only removed reſpectively more diſtant from the Obſerver's Station; and that at leaſt this we may be certain of, that they are all together undoubtedly an Infinity of like Bodies, diſtributed either promiſcuouſly, or in ſome regular Order throughout the mundane Space: And, as *Marino* ſays,

> Reſplendent Sparks of the firſt Fire!
> In which the Beauty we admire,
> And Light of thoſe eternal Rays,
> The uncreated Mind diſplays.

It remains now I think to ſhew, and endeavour to prove, that the Stars are not only light Bodies of the Nature of the Sun, but that they are really ſo many Suns, all performing like Offices of Heat and Gravity, in a regular Order, throughout the viſible Creation, in oppoſition to an Opinion you

* Mr. *Bradley*, Aſtronomer Royal, has, in a great meaſure, proved that the Aberration of the Stars hitherto miſtaken for a Parallax, may ariſe from, and indeed ſeems to be no other than the progreſſive Motion of Light, and Change of Place to the Eye, ariſing from the Earth's annual Motion and Direction.

you have formerly hinted at, of their being in another Senſe of a ſecondary Nature.

All Objects within the ſenſible Sphere of the Sun's Attraction, or Activity, are in ſome meaſure magnified by a good Teleſcope: But the Stars are all placed ſo far without it, that the beſt Glaſſes has no other Effect upon them than making them appear more vivid or lively, but all inate opaque Bodies, reflecting only a borrowed Light from ſome primary one, contrary to this Property, are all obſerved to loſe their Light, in the ſame Proportion, as they are magnified, and through all Glaſſes become more dull than otherwiſe they appear to the naked Eye: And hence we may infer, without any further Evidence, that the Stars are all light Bodies endowed with native Luſtre; and that Bodies, like the known Planets, from the ſame Reaſoning, it is as clear they cannot be, becauſe their Diſtance, though uncertain as to the Truth of the whole, yet ſuch a Part of it as cannot be denied, would render them all in ſuch a Caſe inviſible.

A Proof of this will plainly preſent itſelf, if we conſider the Courſe of the known Comets, who all of them, without Exception, become imperceptible, and intirely diſappear; though moſt of them much bigger than the Earth, or any of the leſſer Planets, long before they arrive at their reſpective Aphelions.

But we are under a kind of Neceſſity to believe them either Suns or Planets, that is either dark or light Bodies; and ſince I have ſhewn the Improbability; nay, I may venture to ſay, the Impoſſibility of their being the firſt, it is natural ſure to conclude, that they muſt be of the laſt Sort; and I am perſuaded, if you but once conſider how ridiculous it is to imagine ſo vaſt a Number of Bodies, all rolling round a Number of inviſible Suns, which muſt otherwiſe be the Caſe, ſince they are ſeen on all Sides of ours, and cannot poſſibly be enlightened by him, or any, how all of them, by any one elſe, you cannot poſſibly have any ſort of Difficulty in this Determination: But that no Arguments may be wanting to enforce your Belief of what is here concluded, it will not be amiſs to put you in Mind of an optical Experiment or two, which cannot fail of convincing you of the vaſt Probability of what is here aſſerted of them; and next to a moral Certainty, demonſtrate the Truth of what ſo many of the beſt Aſtronomers have advanced, as before namely, that the Stars are all, or moſt of them, Suns like ours.

Place any concave Lenſe before your Eye, and you will find all viſible Objects will appear through it, as removed to a much greater Diſtance than they really are at, and reciprocally as much diminiſhed. Now, if you

you look upon one of theſe Glaſſes of a proper Concavity, oppoſed to the Sun or Moon, you will reſpectively have the Appearance of a real Star or Planet, the firſt exhibited by the Body of the Sun, the other by the Moon, and either more or leſs diminiſhed in Proportion to the Surface of the Sphere the Glaſs is ground to.

For Example, a double Concave, or Glaſs of a negative Focus, ground to a Sphere of about three Inches Diameter, will if oppoſed to the Sun's Diſk at a proper Diſtance from the Eye, help you to a very good Idea how the Sun appears to the Planet *Jupiter*; and if a proper Regard be had to the Diſtance of the Planet *Saturn*, a Lenſe ſtill more concave may be formed to give a juſt Idea of the Sun's Appearance to *Saturn*. Again, one much more concave than the former, proportioned to the Orbit of *Mars*, will naturally exhibit the ſolar Body, as ſeen from that Planet.

To the Planet *Venus* and *Mercury*, the Sun appearing much larger than to us at the Earth, to have any tolerable Notion of his varied Phænomena to them, it will be neceſſary to procure Glaſſes of a ſuitable Convexity, ground to reciprocal Concaves, which may eaſily be done to any Focus, ſo as to ſhew how the Sun, naturally appears to the Inhabitants of thoſe two Planets.

The various Appearances of the Planets themſelves to us at the Earth, may alſo well enough be had, if through Glaſſes analagous to their reſpective Diſtance and Magnitude, we look at the Moon, particularly all the Phaſes of *Venus*, and even of *Mercury*, and the Giboſity of *Mars*, *&c.* may be juſtly and beautifully repreſented at different Ages of the Moon, as thoſe Planets appear through the largeſt and beſt Teleſcopes.

This Way you may convince even your Friend * * *, who you tell me has reaſoned all his Senſes uſeleſs, and yet continues ſo great an Atheiſt in Aſtronomy, as not to believe the World turns round upon its Axis, though he gives no better Reaſon for it than that of his not being giddy.

After all theſe Arguments, I hope no new Difficulties will ariſe to retard your Belief, or deprive the Stars of their ſolar Nature, ſo juſtly due to them: This Point gained, the next Thing to be conſidered is, whether all thoſe glorious Bodies, the far greater Part of whom being inviſible to the naked Eye, were made purely and purpoſely for the ſole Uſe of this diminitive World, our little trifling Earth.

——Men, conceited Lords of all,
Walk proudly o'er this pendent Ball,
Fond of their little Spot below,
Nor greater Beings care to know,
But think thoſe Worlds, which deck the Skies,
Were only form'd to pleaſe their Eyes. DUCK.

82 The

The very Suppoſition not only implies a profound Ignorance of the Divine Attributes, but is as impious, and full of Vanity, as it is erroneous and abſurd, and even a Blindneſs ſufficient of itſelf, were there no other Cauſe for it, to introduce Idolatry in the Minds of Mortals, by ſinking the divine Nature ſo near to the human.

It being granted that the Stars are all of the ſame Kind, I think it may be agreed, that what we evince of any one may be allowed to be true of any other, and conſequently of all the reſt. This *Poſtulata* gained, I ſhall next proceed to enquire what the real Uſe and Deſign of ſo many radiant Bodies are, or may be made for.

The Sun we have juſtly reduced to the State of a Star, why then in Reaſon ſhould he have his attendant Planets round him, more than any of the reſt, his undoubted Equals? No Shadow even of a Reaſon can be given for ſuch an Abſurdity.

May we not with the greateſt Confidence imagine, that Nature as juſtly abhors a *Vacuum* in Place, as much as Virtue does in Time? Surely yes: And by ſuppoſing the Infinity of Stars, all centers to as many Syſtems of innumerable Worlds, all alike unknown to us; how naturally do we open to ourſelves a vaſt Field of Probation, and an endleſs Scene of Hope to ground our Expectation of an *ever*-future Happineſs upon, ſuitable to the native Dignity of the awful Mind, which made and comprehends it; and whoſe Works are all as the Buſineſs of an Eternity?

If the Stars were ordained merely for the Uſe of us, why ſo much Extravagance and Oſtentation in their Number, Nature, and Make? For a much leſs Quantity, and ſmaller Bodies, placed nearer to us, would every Way anſwer the vain End we put them to; and beſides, in all Things elſe, Nature is moſt frugal, and takes the neareſt Way, through all her Works, to operate and effect the Will of God. It ſcarce can be reckoned more irrational, to ſuppoſe Animals with Eyes, deſtined to live in eternal Darkneſs, or without Eyes to live in perpetual Day, than to imagine Space illuminated, where there is nothing to be acted upon, or brought to Light; therefore we may juſtly ſuppoſe, that ſo many radiant Bodies were not created barely to enlighten an infinite Void, but to make their much more numerous Attendants viſible; and inſtead of diſcovering a vaſt unbounded deſolate Negation of Beings, diſplay an infinite ſhapeleſs Univerſe, crowded with Myriads of glorious Worlds, all variouſly revolving round them; and which form an Atom, to an indefinite Creation, with an inconceivable Variety of Beings and States, animate and fill the endleſs Orb of Immenſity.

F That

That the ſidereal Planets are not viſible to us, can be no Objection to their actual Exiſtence, and being there, is plain from this; it is well known, that the Stars themſelves, which are their Centeral, and only radiant Bodies, are little more to us at the Earth, than mathematical Points. How ridiculous then is it to expect, that any of their ſmall opaque Attendance, ſhould ever be perceived ſo far as the Earth by us; and beſides, to ſhow the Impoſſibility of ſuch a Diſcovery, we need only conſider, what is, and what is not to be expected, or known in our own home Syſtem. All the Planets in this our ſenſible Region, every Aſtronomer knows, is far from being viſible to one another, in every individual Sphere; for to an Eye at the Orb of *Saturn*, this Earth we live upon, which requires Years to circumſcribe, and Ages to be made acquainted with, and is far from being yet all known, cannot poſſibly from the above Planet be ſeen: And further, ſince *Saturn* and *Jupiter*, two of the moſt material and conſiderable Globes we know of, except the Sun himſelf, are Bodies apparently of the ſame kind, and are obſerved to have each a Number of leſſer Planets moving round them; why may we not expect with equal Certainty and Propriety, that all other Bodies, under the ſame Circumſtances, are in like manner attended; that is, ſeeing the Sun is found to be the Center of a Syſtem of Bodies, all variouſly volving round him? where lies the Improbability of his fellow Luminaries, the Stars, being ſurrounded in like ſort, with more or leſs of ſuch Attendance.

I ſhall offer but one Thing more to your Conſideration in this Affair, and which I am in great Hopes will be ſufficient to make you think theſe natural Suggeſtions a good deal more than probable, and that is this:

The modern Aſtronomers having, in a great meaſure, proved that the Stars are, in all reſpects, vaſt Globes of Fire like our Sun. Let us ſuppoſe a new-created Mind, or thinking Being, in a profound State of Ignorance, with regard to the Nature of all external Objects, but fully endowed with every human Senſe and Force of Reaſon, ſuſpended in Æther, exactly in the midway, betwixt * *Syrius* and the Sun; in which Caſe, both of theſe Luminaries would equally appear much about the Brightneſs of the largeſt of our Planets. Now ſhould ſuch a Being, determined either by Accident or Choice, arrive at this our Syſtem of the Sun, and ſeeing all the planetary Bodies moving round him, I would aſk you what you think he would imagine to be round *Syrius*? Your Anſwer, I think I may venture to ſay, would not be *nothing*; and methinks I already hear you ſay, Why Planets ſuch as ours.

* A Star of the firſt Magnitude in the greater *Dog*, and the moſt neighbouring to our Sun.

PLATE

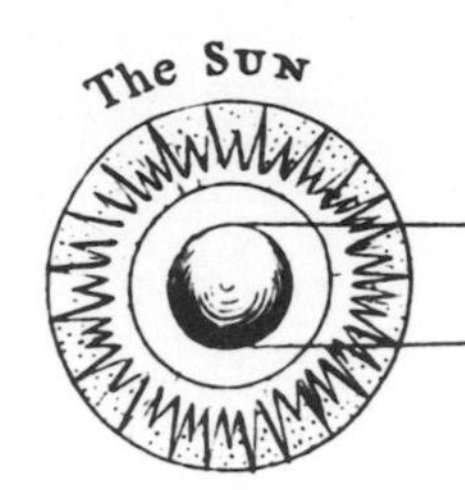

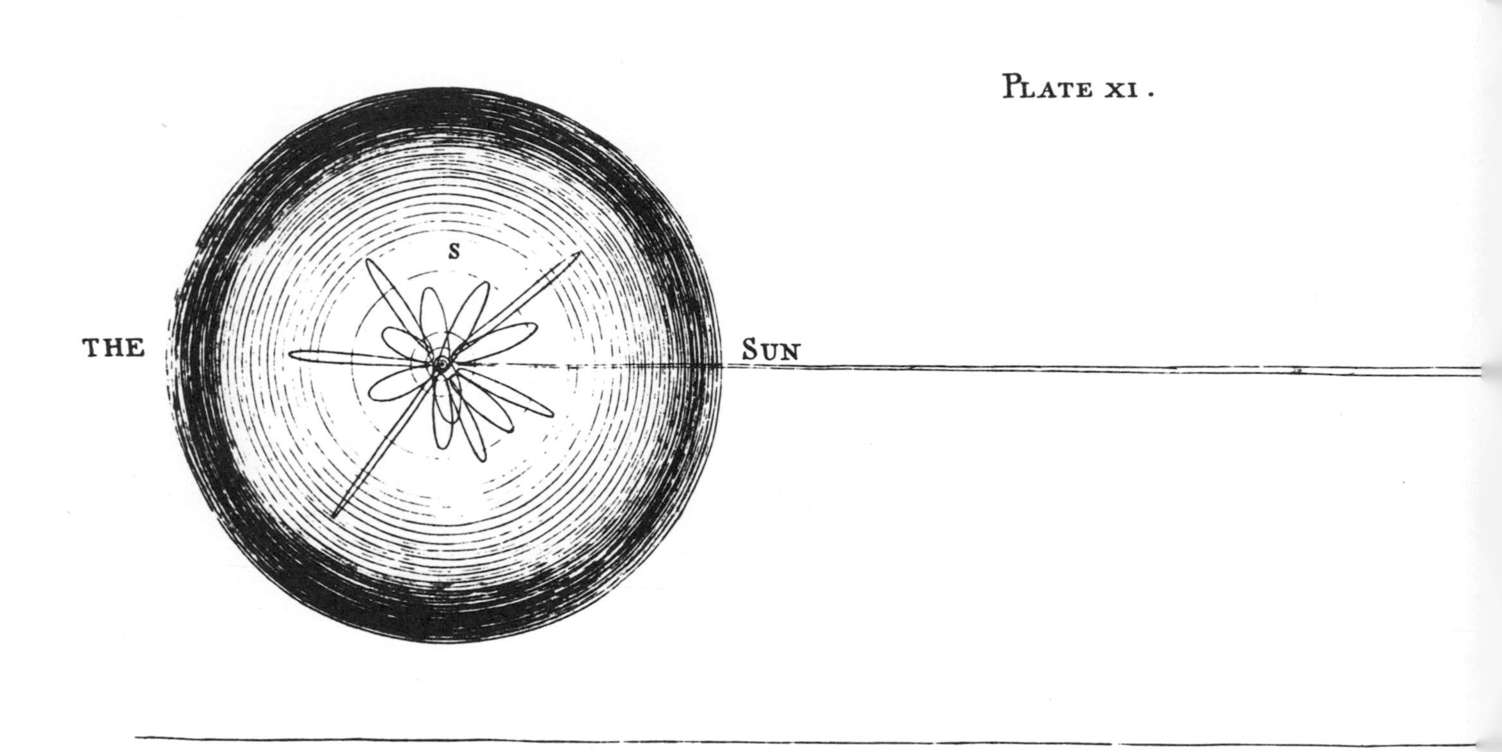

PLATE XI.

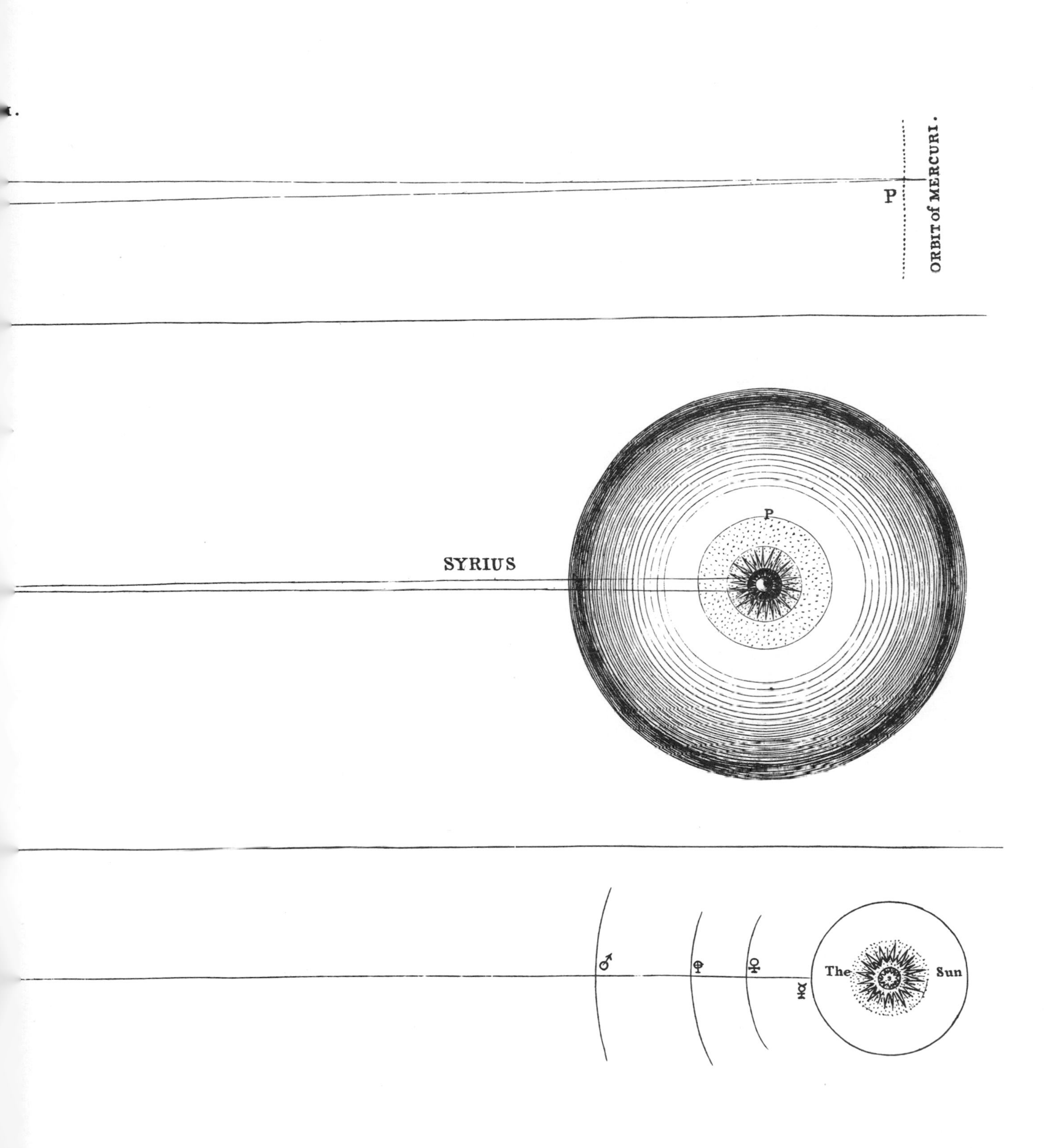

ORBIT of MERCURI.
P
SYRIUS
P
The
Sun

PLATE XI.

Is designed as a geometrical Scale to all the primary Parts of the visible Creation, with regard to the Distance of Orbits compared with the Globe of the Sun; by which at once may be conceived, and justly measured in the Mind, not only the mean Distance of the Planets with regard to one another, but also that of the Comets, and even the comparative Distances of the nearest of the Stars, which will, I guess, greatly help you to form an Idea of the vast Extent of Space necessary to comprehend the whole Creation.

Fig. 1. Is a Radius of the Orbit of *Mercury*, in true Proportion to the Body of the Sun represented at S, shewing at the same time a small Portion of the opaque Planet's Orbit, and the real Length of its Shadow at P.

Fig. 2. Is a Radius of the whole System of the Planets as far as the Orbit of *Saturn* in Proportion to a compleat Orbit of *Mercury*, much less than the former; the former serving as a better known Scale to consider the amazing Distances of the more remote Planets by.

Lastly, *Fig.* 3. Is a Representation of the least possible Distance of *Syrius* and the Sun, proportionable to the Magnitude of the Sphere of our Comets, *&c.* represented at S, whereby it evidently appears, that as all the Planets of *Syrius* must be included within the small Sphere represented in the Center P, none of them could possibly be seen at the Sun, not only by reason of the Smallness of the Angle of Sustension, or Elongation, but also as being lost in the superior Light of *Syrius* himself, in so minute an Orb of Vicinity.

Consequently (as you must perceive) no Arguments can possibly be drawn to deny the Existence of such Bodies, with any Shew of Reason, from their not having been seen by us.

Here I must observe to you, that you cannot consider this Scale of Orbits too much before you look upon Plate XVII.

To conclude, it evidently seems to be the End and Design of Providence, by this visible Variety of Beings, to lift the Minds of Men above this narrow Earth, in Search of that powerful Being upon which we are all so much dependant; and the *Creator*, no doubt, in this vast Display of his Wisdom and Power, designed the amazing Whole, as the adequate Object of every Part, and as such equally open on all Sides, to the penetrating Progress of human Minds, and through the most extensive Faculty of Sense, the *Sight*, to draw our Reason and Understanding by Degrees, from finite Objects into Infinity; and as the last Result of celestial Contemplations place within our Reach, a certain Evidence of a future State, *and the manifest Mansions of Rewards and Punishments, suited no doubt most equitably to all Degrees of Virtue, and to every Vice.*

"When I consider (says Mr. *Addison*, speaking as having taken particular "notice of a fine Evening) that infinite Host of Stars, or to speak more "philosophically of Suns, which were then shining upon me, with those "innumerable Sets of Planets or Worlds, which were then moving round "their respective Suns; when I still enlarge the Idea, and supposed ano- "ther Heaven of Suns and Worlds rising still above this which we dis- "covered; and these still enlightened by a superior Firmament of Lu- "minaries, which are planted at so great a Distance, that they may ap- "pear to the Inhabitants of the former as the Stars do to us; in short, "whilst I pursued this Thought, I could not but reflect on that little "insignificant Figure which I myself bore amongst the Immensity of "God's Works:" This Reflection, I judge, as you are an Admirer of the Author, you will not look upon as impertinent in this Place, especially as it must enforce what I have endeavoured to shew you, namely, the Reasonableness of a Plurality of sidereal Systems, and their Multiplicity of Worlds; which, if you are yet in Doubt of, I hope you will at least forgive so well designed an Attempt with your usual Candour.

I am now prepared to proceed in the chief Design of this Undertaking, which is to solve the Phænomena of the *Via Lactea*; and propose in my next to answer more fully your farther Request.

I am, &c.

LETTER

LETTER THE FIFTH.

Of the Order, Distance, and Multiplicity of the Stars, the Via Lactea, *and Extent of the visible Creation.*

SIR,

WE are told, and, if I remember right, it is also your Opinion, that three of the finest Sights in Nature, are a rising Sun at Sea, a verdant Landskip with a Rainbow, and a clear Star-light Evening: All of which I have myself often observed with vast Delight and Pleasure. The first I have frequently beheld, and always with an agreeable Surprize; the second I have as often taken notice of, with no small Degree of Admiration; but the last I shall never look up to without an Astonishment, even mixed with a kind of Rapture. The Night you last left us, this admirable Scene was in its full Beauty; and, as *Milton* says,

> Silence was pleas'd: now glow'd the Firmament
> With living Saphirs; *Hesperus* that led
> The starry Host rode brightest.———

PL 4.604

I found it was impossible to look long upon this stupendious Scene, so full of amazing Objects, and particularly the *Via Lactea*, which (the Moon being absent) was then in great Perfection, without being put in Mind of my Task. This surprizing Zone of Light being the chief Object I have undertaken to treat of and demonstrate.

This amazing Phænomenon which have been the Occasion of so many *Fables*, idle Romances, and ridiculous Opinions amongst the Antients, still continues to be unaccounted for, and even in an Age vain enough to boast Astronomy in its utmost Perfection.

What will you say, if I tell you, it is my Belief we are so far from the real Summit of the Science, that we scarce yet know the Rudiments of what may be expected from it. This luminous Circle has often engrossed my Thoughts, and of late has taken up all my idle Hours; and I am now in great

great Hopes I have not only at laſt found out the real Cauſe of it, but alſo by the ſame Hypotheſis, which ſolves this Appearance, ſhall be able to demonſtrate a much more rational Theory of the Creation than hitherto has been any where advanced, and at the ſame Time give you an intire new Idea of the Univerſe, or infinite Syſtem of Things. This moſt ſurprizing Zone of Light, which have employed ſucceſſively for many Ages paſt, the wiſeſt Heads amongſt the Antients, to no other Purpoſe than barely to deſcribe it; we find to be a perfect Circle, and nearly biſecting the celeſtial Sphere, but very irregular in Breadth and Brightneſs, and in many Places divided into double Streams.

* The principal Part of it runs through the *Eagle*, the *Swan*, *Caſſiopea*, *Perſeus*, and *Auriga*, and continues its Courſe by the Head of *Monoceros*, along by the greater *Dog* through the Ship, and underneath the *Centaur*'s *Feet*, till having paſſed the *Alter*, the *Scorpion's Tail*, and the Bow of *Aquarius*, it ends at laſt where it begun.

PLATE XII, and XIII.

Repreſents the two Hemiſpheres, where its true Tract is diſtinguiſhed amongſt the principal Stars, and may eaſily be conceived by them to circumſcribe and biſect the whole Heavens.

This is that Phænomena I am about to explain and account for; but before I proceed farther, I judge it will be no *improper Precognita*, to give you the Thoughts of the Antients upon it; the Relation perhaps may require ſome Patience; but I gueſs, that after reading ſuch wild and extravagant Notions concerning it, you will naturally judge more favourably of the Conjectures of the Moderns upon it, and particularly of what is concluded in the ſucceeding Pages.

Theophraſtus

* ——— Carried toward the oppoſed *Bears*,
Its Courſe cloſe by the *Artick* Circle ſteers,
And by inverted *Caſſiopea* tends;
Thence by the *Swan* obliquely it deſcends
The Summer Tropick, and *Jove's* Bird divides;
Then croſs the Equator, and the Zodiack glides
'Twixt *Scorpio*'s burning Tail, and the left Part
Of *Sagitarius*, near the fiery Dart;
Then by the other *Centaur*'s Legs and Feet,
Winding remounts the Skies (again to meet)
By *Argos*' Topſail, and Heav'ns middle Sphere,
Paſſing the *Twins*, t' o'ertake the Charioteer;
Thence *Caſſiopea* ſeeking thee does run,
O're *Perſeus* Head, and Ends where it begun.

SHER. MANILIUS.

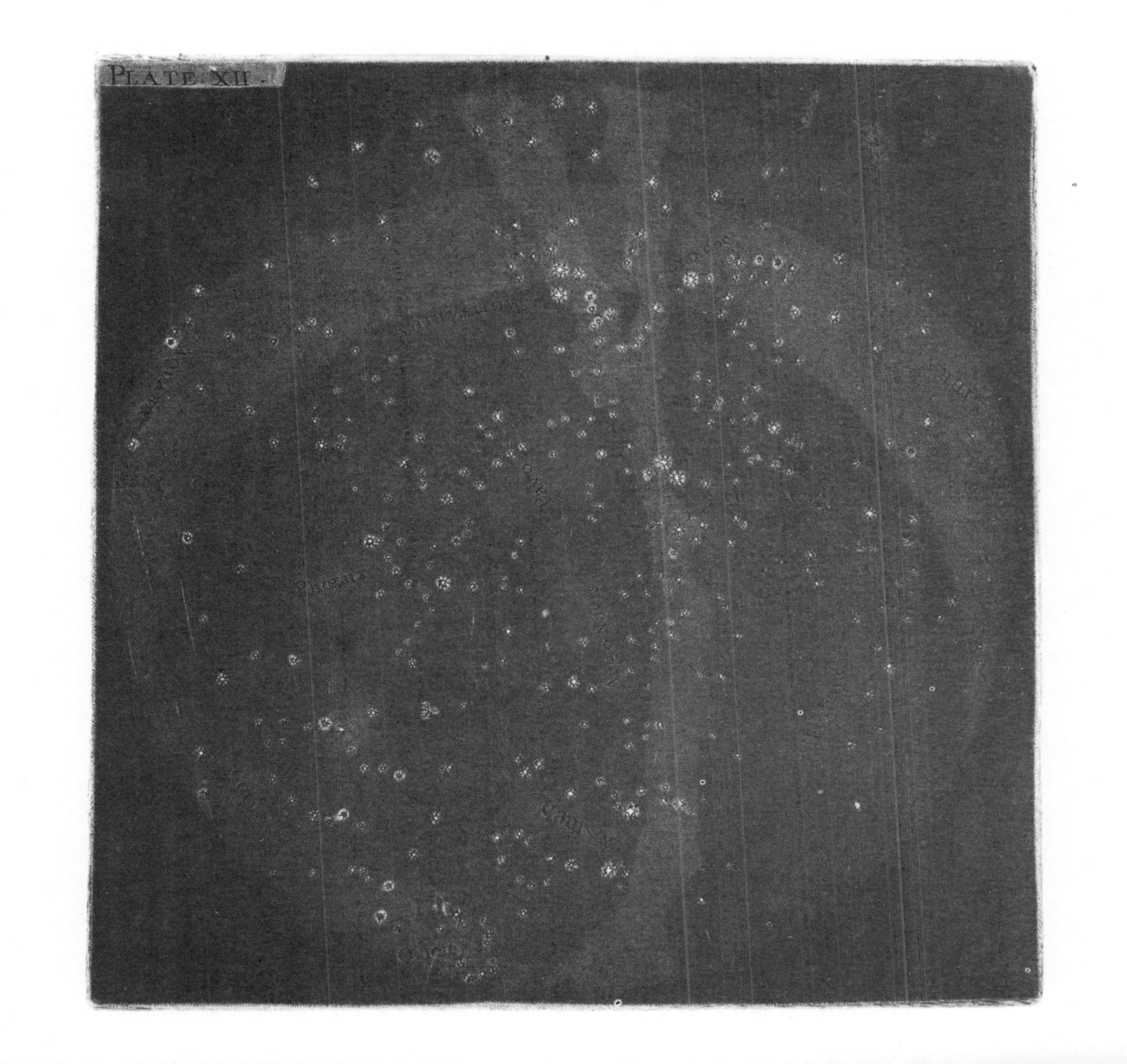
PLATE XII.

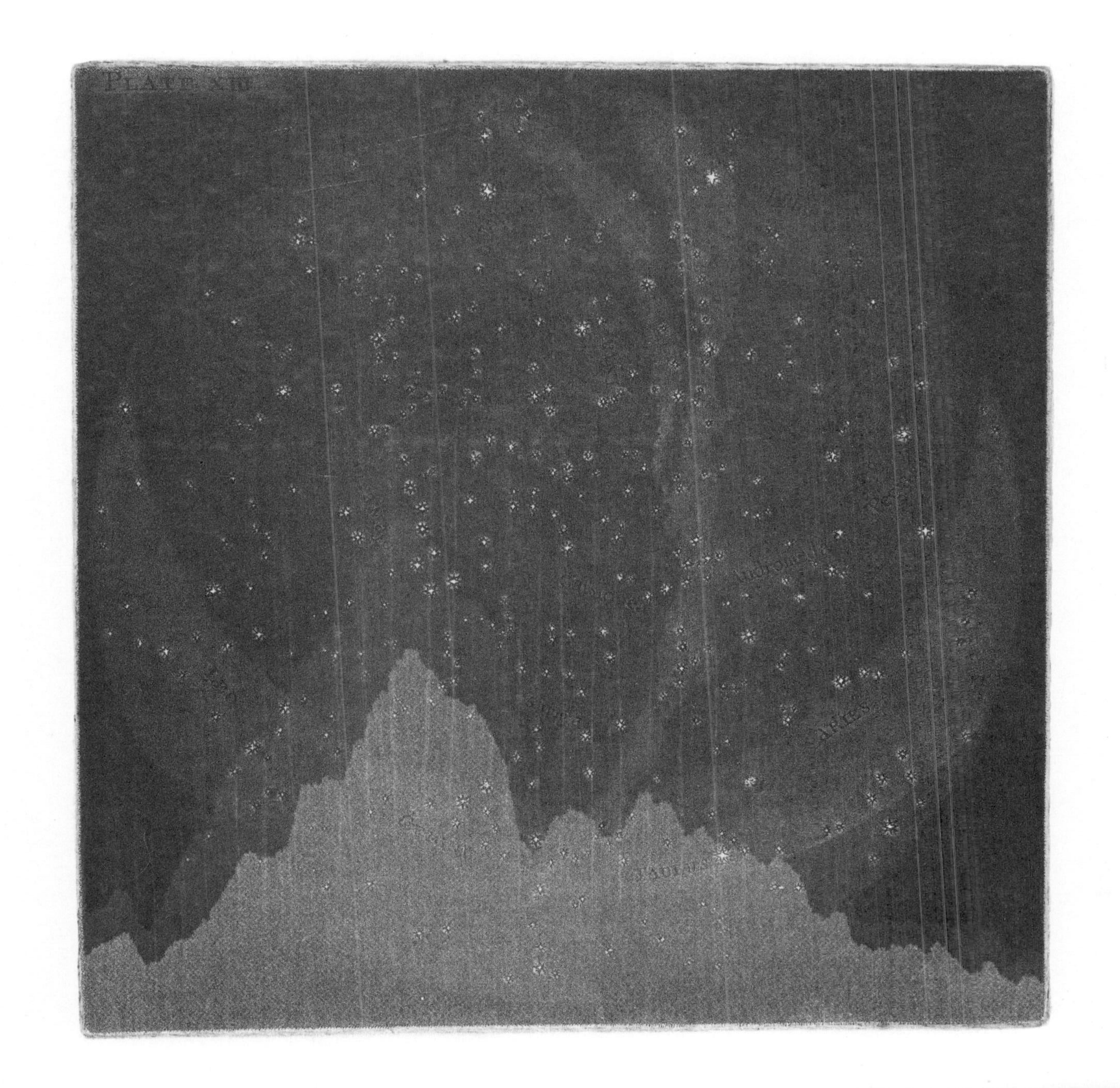
Plate XII
Andromeda
Aries

Theophrastus * was of Opinion, that the Hemispheres, which, by many of the Antients were imagined to be solid, was joined together here; and that this was the soldering of the two Parts into one. † DIODORUS thought it celestial Fire, of a dense and compact Nature, seen through the Clifts or Cracks of the parting Hemisphere: But as *Manilius* says,

> Astonishment must sure their Senses reach,
> To see the World's wide Wound, and Heav'n's eternal Breach.

OENOPIDES ‖ believed it the ancient Way of the Sun, till frighted at the bloody Banquet of *Thyestis.* ** ERATOSTHENES supposed it *Juno*'s Milk, spilt whilst giving Suck to *Hercules.* ‡‡ PLUTARCH makes it the Effect of *Phaeton*'s confused Erratication; but I think it is plain †† OVID judged them to be Stars, and the ancient *Ethnicks* believed them to be the blissful Seats of valiant and heroic Souls.

> ——Valiant Souls, freed from corporeal Gives,
> Thither repair, and lead æthereal Lives.

MANILIUS.

* *Macrobius*, lib. i. cap. 15.
Or meets Heaven here! and this white Cloud appears
The Cement of the close-wedg'd Hemispheres!

† The sacred Causes human Breasts enquire,
Whether the heavenly Segments there retire,
(The whole Mass shrinking, and the parting Fame
Thro' cleaving Chinks admits the stranger Flame.

‖ Or seems that old Opinion of more Sway,
That the Sun's Horses here once run astray,
And a new Path mark'd in their straggling Flight,
Of scorching Skies, and Stars adusted Light.

** Nor must that gentle Rumour be supprest,
How Milk once flowing from fair *Juno*'s Breast
Stain'd the celestial Pavement, from whence came
This milky Path, its Cause shewn in its Name.

‡‡ When from the hurried Chariot Light'ning fled,
And scatter'd blazes all the Skies o'erspread;
By whose Approach new Stars enkindled were,
Which still as Marks of that sad Chance appear.

MANILIUS.

†† A Way there is in Heaven's expanded Plain
Which when the Skies are clear, is seen below,
And Mortals by the Name of *Milky*, know,
The Ground-work is of Stars ----

Ovid's Met. lib. i.

But

But * DEMOCRITUS long ago believed them to be an infinite Number of ſmall Stars; and ſuch of late Years they have been diſcovered to be, firſt by *Gallaleo*, next by *Keplar*, and now confirmed by all modern Aſtronomers, who have ever had an Opportunity of ſeeing them through a good Teleſcope.

PLATE XIV.

Is from an Obſervation I made myſelf, of a bright Part of this Zone near the Feet of *Antinous*; which, (by a Miſtake of the Engraver) is, as it appears through a Tube of two convex Glaſſes. I ſaw it through a very good Reflector, and formed the Plan by a Combination of Triangles.

Milton takes notice of this Zone in a moſt beautiful Manner, where he deſcribes the Creator's Return from his ſix Day's Work to Heaven, he introduces it as a Simile to expreſs his Idea of the eternal Way, or Road to the celeſtial Manſions.

> —— A broad and ample Road, whoſe Duſt is Gold
> And Pavement Stars, as Stars to thee appear,
> Seen in the *Galaxie*, that Milky Way,
> Which nightly as a circling Zone thou ſeeſt
> Powder'd with Stars.

But to infer from their Appearance only, that they are really Stars, without conſidering their Nature and Diſtance; and that nothing but Stars could poſſibly produce ſuch an Effect, may perhaps be aſſuming too much, when we have nothing but the bare Credit of the *Belgic* Glaſſes to ſupport our Conjectures; and although this may be ſufficient for any Mathematician, yet for your greater Satisfaction, I have thought proper to give two or three more evincing Arguments, to confirm theſe important Diſcoveries. *Democritus*, as I have ſaid before, believed them to be Stars long before Aſtronomy reaped any Benefit from the improved Sciences of Optics; and ſaw, as we may ſay, through the Eye of Reaſon, full as far into Infinity as the moſt able Aſtronomers in more advantageous Times have done ſince, even aſſiſted with their beſt Glaſſes: And his Conjectures are almoſt as old as the philolaic Syſtem of the Planets itſelf; the Conſtruction of which, though attempted by many, none have ever yet been able to confute.

The Light which naturally flows from this Crowd of radiant Bodies is mixt and confuſed, chiefly occaſioned by the Agitation of our Atmoſphere, and from a Union of their Rays of Light, by a too near Proximity of their Beams, altogether they appear like a River of Milk, but more of a pelucid Nature, running all round the ſtarry Regions.

For

* *Plutarch (in Placitis Philoſoph.)*

PLATE XIV.

For in the azure Skies its candid Way
Shines like the dawning Morn, or closing Day.

There are also many more such luminous Spaces to be found in the Heavens of the same Nature with these, which we know to be Stars; in particular the *Nebulæ*, or cloudy Star in the *Præsepe* of 36; a cloudy Star in *Orion* of 21; * a cloudy † Knot not far from this in the same Asterism of 80; in one Degree of the same Constellation 500, and in the whole Form above ‡ 2000. All of which are great Confirmations of the Truth of our Assertion, *i. e.* that this Zone of Light proceeds from an infinite Number of small Stars. Here it will not be amiss to observe, that it has been conjectured, and is strongly suspected, that a proper Number of Rays, meeting from different Directions, become Flame; and that hence it may prove not the Sun's real Body which we daily see, but only his inflamed Atmosphere. I begin to be of Opinion, and I think not without Reason, that the true Magnitude of the Sun is not near what the modern Astronomers have made it; and that it may not possibly be much above two Thirds of what it appears to us; I don't mean that this Expansion of the solar Flame is any Part of that dilated Light mentioned by Sir *Isaac Newton*, and conceived to be round all light Bodies in general; but you may consider it as not much differing from it, not of an unlike Nature, only greater in Degree, and peculiar to the Sun and Stars, who are all, as has been before in a manner demonstrated to be actually Globes of Fire.

This, tho' I presume to call it at present only meer Hypothesis, will in a great measure account for the excessive Changes in the Constitution of our Air and Atmosphere, which we often find very unnatural to the Season; also be a Means perhaps of reconciling the vast Disproportion so very remarkable betwixt the Sun and the lesser Planets, and many other Circumstances in the System of no small Consequence in Astronomy: One of which Particulars you have frequently expressed a great Mistrust and Disapprobation of, as suspectng some kind of a Fallacy in the Computation; and the other is Matter of general Complaint, being by many attributed to a Change in the Direction of the Earth's Axis ‖; and by some, especially the Vulgar, to too near an Approximation of the Earth to some one of the celestial Bodies. But all this will very naturally be accounted for by the Levity, or expanding Quality of the Sun's circumambient

* Vide *Galilæo*.
† Betwixt the Sword and Girdle of *Orion*.
‡ Vide *Reitha*.
‖ Which, through Ignorance of the true Case, is commonly called a Shock, a Brush, or Shove.

Flame, or Atmoſphere; and hence, according to its various State, being more condenſed, or rare, we may have Heat or Cold in the greateſt Extream, and alternately ſo, in a perpetual Viciſſitude.

The Truth of this Doctrine will evidently appear from the Obſervations of the Sun's Diameter through the Year 1660, by the indefatigable *Mouton:* And, I muſt own, I am not a little ſurprized to find that no Concluſions have been drawn from them of this Kind. I am perſwaded, if you once compare thoſe Numbers, you will be very far from thinking this an improbable Suggeſtion. But this Digreſſion has led me a little too far from the *Via Lactea*, and too near home again; I muſt now think of returning to the Stars, and my next Endeavours muſt be to give you ſome Idea of the Number of them. Through very good Teleſcopes there have been diſcovered in many Parts of this enlightened Space, and even out of it, ſeveral thouſand Stars in the Compaſs of one ſquare Degree; in particular near the Sword of *Perſeus*, and in the Conſtellations of * *Taurus* and *Orion*.

PLATE XV.

Repreſents the *Pleides*, a well known Knot of Stars in the Sign *Taurus*, as they appeared to me thro' a one Foot reflecting Teleſcope: And *Plate* XVI. is a View of the *Perſides*, another ſurprizing Knot of Stars in the Conſtellation *Perſeus*, exactly as they appear through a Tube of two convex Glaſſes. There are alſo other luminous Spaces in the ſtarry Regions, not unlike the Milky Way, which I have had no Opportunity of obſerving; ſuch as the *Nebeculæ*, near the South Pole, called by the Seamen *Magellanic* Clouds; and which likewiſe viewed through Teleſcopes, preſent us with little *Nebulæ*, and ſmall Stars interperſed: One of theſe Kind is ſituated between *Hydrus* and *Dorado*; and another, ſomething leſs than this, betwixt *Hydrus* and the *Toucan*.

Now admitting the Breadth of the *Via Lactea* to be at a Mean but nine Degrees, and ſuppoſing only twelve hundred Stars in every ſquare Degree, there will be nearly in the whole orbicular Area 3,888,000 Stars, and all theſe in a very minute Portion of the great Expanſe of Heaven. What! a vaſt Idea of endleſs Beings muſt this produce and generate in our Minds; and when we conſider them all as flaming Suns, Progenitors, and *Primum Mobiles* of a ſtill much greater Number of peopled Worlds, what leſs than an Infinity can circumſcribe them, leſs than an Eternity comprehend them,

* *Galilæo* in one cloudy Star of this Conſtellation, diſcovered no leſs than twenty-one, and in that of the *Præſepe* thirty-ſix.

or

PLATE. XV.

PLATE XVI.

or leſs than Omnipotence produce and ſupport them, and where can our Wonder ceaſe?

In this Place perhaps I ought not to paſs over the aſtoniſhing Phenomenon of ſeveral new Stars, *&c.* which have frequently appeared, and ſoon again vaniſhed, in the ſame Point of the Heavens. But as the Buſineſs of this Theory is rather to ſolve the general, than any particula Phænomenon, I ſhall only here by way of Note ſubjoin a Table of ſuch as has been regularly obſerved, and by whom they were firſt diſcovered.

A Table of ſeveral new Stars, Nebulæ, *and double Stars,* &c.

Nomina Stellarum.	*Obſervationum.*
Septima Pleiadum	Loſt after the burning of *Troy*, but now returned; ſee *Ricciol.s.*
A new Star appeared in *Caſſiopea*, nearly in the ſame Place with that of 1572.	*Anno Dom.* 945, bright as *Jupiter*; ſee *Ricciolus.*
The new Star in *Caſſiopea*'s Chair.	Bright as *Venus*, from *November* 1572 to *March* 1574.
A new Star in *Collo Ceti.*	Of the 3d Magnitude, is ſaid to have appear'd periodically, ſeven Times in ſix Years, *i. e.* every three hundred and thirteen Days: It was firſt obſerved in *Auguſt* 1596, for two Months, by *D. Fabricius*
A new Star in the Swan's Neck,	Obſerved by *Kepler* in 1600, of the third Magnitude, till the Year 1659; then gradually decreaſing; in 1661 it diſappeared; in 1666 it became viſible again, and is yet to be ſeen of the ſixth Magnitude.
A new Star in the Right Foot of *Serpentarius*,	Bright as *Venus* from *October* 1604 to *October* 1605: ſee *Kepler.*
A new Star in *Andromeda*'s Girdle,	Seen by *Simon Marius* and *Fabricius*, *Anno* 1612.
A new Star in *Antinous*,	Seen by *Juſtus Byrgius.*
A new Star ſeen in the Whale,	In 1638, by *John Procyclides Holuarda*, of the third Magnitude, which diſappeared periodically, every three hundred and thirty Days.
A new Star in the Fox's Head,	Of the third Magnitude, ſeen by *Hevelius* in *July* 1670, and till *Auguſt* 1671, alſo from *March* 1672 to *September* 1672
A new Star in the Swan's Neck.	This appear'd periodically every four hundred and four Days, and about ſix Months at a Time; it was ſeen at its brighteſt, *September* 10, 1714.

Of the Nebulæ, *or Cloudy Stars.*

Nebuloſe in *Orion*'s Sword.	
Nebuloſe in *Andromeda*'s Girdle.	
Nebuloſe in the Bow of *Sagitarius*,	Small, but very luminous.
Nebuloſe in *Centaurus*,	Never ſeen in *England.*
A *Nebuloſe* preceding the right Foot of *Antinous*,	Obſcure, but with a Star in the Middle of it.
Nebulæ in *Dorſo Herculis*,	Diſcovered by Dr. *Hally*

Beſides the *Nebulæ*, and new Stars, it appears from the ancient Catalogues of *Hevelius*, &c. that ſome of the old ones have intirely vaniſhed; in particular, one in the left Thigh of *Aquarius*, the contiguous one preceding in the Tail of *Capricorn*; the ſecond on the Belly of the Whale; the firſt of the unformed ones after the Scales of *Libra*, and ſeveral others. Many of the Stars alſo appear to be double, as the firſt Star of *Aries* and *Caſtor*; others triple, as one in the *Pleiades*; and the middle one in *Orion*'s Sabre; and others again, quadruple, *&c.*

I would

I would now willingly help you to conceive the indefinite mutual Diſtance of the Stars, in order to give you ſome ſmall Notion of the Immenſity of Space; but as this will be a Taſk merely conjectural, I ſhall only deſire you to believe it as far as your Reaſon will carry you, ſafely ſupported by an obvious Probability.

Perhaps it may be neceſſary here to acquaint you, that all the Stars are ſo far apparently of different Magnitudes, that no two of them are to be found in the whole Heavens exactly the ſame, either in Bigneſs or Brightneſs *. The largeſt we have ſufficient Reaſon to believe is the neareſt to us; the next in Bigneſs and Brightneſs more remote; and ſo on to the leaſt we ſee, which we judge to be the moſt remote of all.

The firſt Degree, or that of the largeſt Magnitude, we give to SYRIUS, the ſecond to ARCTURUS, the third to ALDEBARAN, the fourth to LYRA, the fifth to CAPELLA, the ſixth to REGULUS, the ſeventh to RIGEL, the eighth to FOMAHAUNT, and the ninth to ANTARUS: Theſe are all ſaid to be of the firſt Claſs; and beſides which, there are at leaſt, within the Reach of our lateſt improved Opticks, nine more Denominations within the Radius of the viſible Creation.

Now, by the certain Return of the Comets, which we find are all governed by the Laws of this Syſtem, and ſuppoſed to be undiſturbed by any of the others, we cannot avoid concluding, if we conſider them at all to the Purpoſe, that the neareſt Stars cannot be leſs diſtant than twice the Radius of the greateſt Orbit belonging to the Sun. Moſt Mathematicians think this a great deal too near, as it muſt of courſe make all the Syſtems join, as in Contact; and I think we may ſafely add, to ſeparate their Spheres of Attraction, at leaſt one Half of this Diſtance more, which will make in the Whole about four hundred and twenty Semi-orbits of the Earth, or 33,600,000,000 Miles. This even the ingenious Mr. *Huygins* endeavours to prove ſtill much too little, and his Arguments are ſuch as cannot eaſily be refuted. His Principle is grounded upon the known Laws of Analogy, as conſidered in the Proportion of light Surfaces, and is as follows. Having reduced the Sun's Diſk to the Appearance of the Star SYRIUS, by the Help of a ſmall Hole at the End of his Teleſcope, and comparing this Part of his Surface to the whole Diſk of the Sun, he infers that the Stars Diſtance to that of the Sun muſt be as 27,664 to 1. Hence *Syrius* from us will be nearly (avoiding Units) 2,213,120,000,000 Miles: But this I take to be as much too large as the former is too little; yet, as

* A very little Knowledge in Opticks will render this indiſputable, and has been in a great meaſure demonſtrated before; 1. in the Great Dog; 2. in Bootes; 3. in the Bull; 4. in the Harp of *Apollo*; 5. in *Auriga*; 6. in the Lion; 7. in *Orion*; 8. in the Southern Fiſh; 9. at the End of *Erridanus*.

 Mr.

Mr. *Bradley* has, with ſome Shew of Reaſon, baniſhed all the Stars out of the Sphere of Parallax, the laſt is the only Method we can poſſibly make uſe of with any kind of Confidence; and Sir *Iſaac Newton* endeavours to recommend it with great Force of Argument, as the only probable Means by which we can give any tolerable Gueſs at theſe immenſe Meaſurements of Space.

To moderate the Matter then if you pleaſe, allow me but to make uſe of a Mean betwixt the two fore-mentioned Numbers; and we may take it for granted, a Diſtance ſufficiently exact, to ſuit all our Wants in the preſent Caſe, namely, to give a very tolerable Idea of the Extent of the viſible Creation, which is all I propoſe in this Place to attempt; but I mean to be much more exact in another.

Now as the Diſtance from the Sun to the Earth is ſo ſmall in Proportion to the Diſtance of the Stars from us, and from one another, we may very well conſider the Sun as the Center of our Station, or Poſition in the general Syſtem or Frame of Nature. And as the Stars are very viſible thro' good Teleſcopes, to the ninth or tenth Magnitude, if we multiply the primary Diſtance of *Syrius*, or of any other of his Claſs, by this Number of common intermediate Spaces, the Product will be equal to the Radius of the viſible Creation to the ſolar Eye; which, by this Rule, you will find in capital Numbers to be * nearly 6,000,000,000,000 Miles, taking in a Star of the ſixth Magnitude, and to a Star of the ninth, 9,000,000,000,000 Miles: But this Computation ſuppoſes a mean common Diſtance of the Stars in a ſort of Syzygia, or Direction of a Right Line, which is not the real Caſe; for the Stars cannot be ſuppoſed to diminiſh in a proportional Magnitude by any mathematical *Ratio*, but by ſome geometrical, or rather muſical one; for Inſtance, if the Diſtance of a firſt be 3, that of a ſecond ſhould be about 5, and of a proportional Third 8,333, *&c. ad infinitum:* But as their true proportional Diſtance is unknown, the above will be ſufficient for our preſent Purpoſe; which is only to ſhew, without Exaggeration, the Space we now are truly ſenſible of.

This I have here conſidered more extenſively, to obviate all Objections that you may make to the Probability of the general Motion of the Stars, by ſhewing no Difficulty can poſſibly ariſe from their apparent Proximity, Number, or irregular Diſtribution: Their Diſtances being ſo immenſely large, no Diſorder or Confuſion can be ſuppoſed in any Direction of them, or Motion whatever. The greateſt Diſtance of the Planets, which all move undiſturbed round the Sun, is about three hundred and fifty-three Million of Miles: But the leaſt Diſtance of one Star from another, is upwards

* If the Diſtance of the Sun and Earth is found too much, which I muſt own I have a violent Suſpicion of, theſe Numbers muſt be reduced in like Proportion.

upwards of two thousand eight hundred and thirty-two Times that Distance, or one Million of Millions of Miles: And as no sensible Disorder can be observed amongst the solar Planets, what Reason have we to suppose any can be occasioned amongst the Stars, or that a general Motion of these primary Luminaries round a common Center, should be any way irrational, or unnatural?

What an amazing Scene does this display to us! what inconceivable Vastness and Magnificence of Power does such a Frame unfold! Suns crowding upon Suns, to our weak Sense, indefinitely distant from each other; and Miriads of Miriads of Mansions, like our own, peopling Infinity, all subject to the same Creator's Will; a Universe of Worlds, all deck'd with Mountains, Lakes, and Seas, Herbs, Animals, and Rivers, Rocks, Caves, and Trees; and all the Produce of indulgent Wisdom, to chear Infinity with endless Beings, to whom his Omnipotence may give a variegated eternal Life.

The astonishing Distance of the starry Mansions undoubtedly was design'd to answer some wise End: One Consequence is this, and probably is not without its Use: To every Planet of the same System, the same sidereal Face of Heaven appears without the least Degree of Change; and as the remotest Regions upon Earth see the same Moon and Planets, so also the Inhabitants of the most distant Planets in ours, or in any other System, see the same Forms and Order of the Stars in common with the rest. The whole Sphere of Heaven being common and unchangeable through all their various Revolutions.

Thus those (the People) in the Planet *Venus* will see the Constellation of *Orion* just as we do, and the People in the Planet *Saturn*, much farther still removed, alike will view this Constellation in all respects the same; here then, (in the System of the Sun) the Eye removed from us must only hope to find a new Earth surrounded with the same sort of Sky: But Beings in another System, behold not only a new Heaven above, but also new Earths below; and all the Frame of Nature to them puts on a new Dress, new Signs, new Seasons, and new Planets roll, and a new Sun renews the Day.

The Heathen Fables here are all erased with all the Immortality of their vain earthly Gods and Heroes; *Perseus* and *Alcides* are no more, and both the *Bears* are vanished; the *Pleiads* and the *Hyads* join, and shining *Leo*, though boasting two Stars of the first Magnitude with us, there no where can be found, lost in the common undistinguished Herd. But still Astronomy will exist, and new-framed Forms may fill the varied Scene.

Perhaps you may expect that I should here give you my Conjectures of what sort of Beings may be supposed to reside in the *Ens Primum*, or *Sedes Beatorum*

Beatorum of the known Univerſe, whether mortal, immortal, or Creatures partaking in ſome Degree of the Properties of both; as ſuch may be conceiv'd to change their Natures and States, without a total Diſſolution of their Senſes by Death: And farther, it may poſſibly be judged unpardonable in me not to point out every bleſſed Abode, ſuited to the Virtues, and all the various States an immortal Soul may be tranſlated to; but this is a Taſk above the human Capacity, or is the pure Province of Religion alone; the Buſineſs of a Revelation rather than Reaſon to diſcover. Beſides, it is enough for the preſent Purpoſe, to prove, that Miriads of celeſtial Manſions, are to be diſcovered within our finite View, and by a kind of ocular Revelation, which viſibly extends the human Proſpect, as it were, far beyond the Grave. It matters not whether a Race of Heroes fill theſe Worlds, or a Tribe of happy Lovers people thoſe; whether a Peaſant in the Realms of *Orion* ſhall ever become a Prince in the Regions of *Arcturus*, or a Patriarch in *Procion*, a Prophet in the *Precepæ*. Not to mention all the Stages human Nature may, or have been deſtined to in any one World, as believ'd by the ancient Philoſophers, beſides the final Coalition of all Beings much more naturally to be expected in the *Sedes Beatorum*.

I ſay, whatever our Caſe may be with regard to theſe *Queries* and Futurity, the Plan and Principles of this Theory will not be at all changed by it, ſince what it is chiefly founded upon may be clearly demonſtrated, ſo clearly and inconteſtably, that, with the Reverend Dr. *Young*, we may juſtly conclude,

Devotion! Daughter of Aſtronomy!

and affirm with him alſo, That,

An indevout Aſtronomer is mad.

But I find what I at firſt propoſed will prove too long for this Letter. However, I will endeavour to reward your Patience in my next, and continue, *&c.*

LETTER

LETTER THE SIXTH.

Of General Motion amongſt the Stars, the Plurality of Syſtems, and Innumerability of Worlds.

SIR,

SINCE my laſt, you'll find by this, ſpeaking in the Stile of *Kercher*, that I have been very far from home, round almoſt the viſible Creation. I have indeed applied myſelf very cloſely to tranſcribe my Thoughts to you upon the old Subject the *Milky Way*, which my former Letter left imperfected. To return then to the Theory of the Stars, and that yet unreconciled Phænomenon; let us reaſon a little upon the viſible Order of the Stars in general, and ſee what Concluſions can be drawn from what every Aſtronomer knows of them, and cannot be diſputed.

Firſt then, that the Stars are not infinitely diſperſed and diſtributed in a promiſcuous Manner throughout all the mundane Space, without Order or Deſign, is evident beyond a Doubt from this vaſt collective Body of Light, ſince no ſuch Phænomenon could poſſibly be produced by Chance, or exhibited without a deſigned Diſpoſition of its conſtituent Bodies.

If any regular Order of the Stars then can be demonſtrated that will naturally prove this Phænomenon to be no other than a certain Effect ariſing from the Obſerver's Situation, I think you muſt of courſe grant ſuch a Solution at leaſt rational, if not the Truth; and this is what I propoſe by my new Theory.

To a Spectator placed in an indefinite Space, all very remote Objects appear to be equally diſtant from the Eye; and if we judge of the *Via Lactea* from Phænomena only, we muſt of courſe conclude it a vaſt Ring of Stars, ſcattered promiſcuouſly round the celeſtial Regions in the Direction of a perfect Circle.

But when we conſider the explanick Poſition of many other Stars, all of the ſame Nature, and not leſs numerous, together forming the great Sphere of Heaven, we generally find ourſelves quite at a Loſs how to reconcileth e two apparent Claſſes; and I know none who have ever been ſucceſsful enough to reduce them to any one general Order.

You'll

You'll ſay probably how ſhall we make this chaoſic Diſpoſition of the primary Luminaries agree with the ſecondary Laws, and the juſt Harmony obſerved in the third * Creation, *&c.*

The Work now you ſee is undertaken, and chiefly at your own Requeſt, therefore I have a Right to expect you'll be very indulgent to the Author, and paſs over all his Faults, and allow him free Argument in Purſuit of theſe important Truths, which will in the End open perhaps a much wider Field of Contemplation to us, than at firſt could be ſuppoſed to be intended by the *Geneſis* of *Moſes*.

That Deſcription of the Beginning of Nature is not without its Beauty and Nobleneſs, ſuitable to the Dignity both of the Author and Subject. But ſhould we even in this knowing Age of the World pretend to account for the Original of Things, as *Moſes* to ſupport his believed divine Legation, was obliged in ſome meaſure to do, we ſhould ſoon be reduced to talk in the ſame Stile, and perhaps with leſs Probability, than then at leaſt appeared in his elegant Account of the Origin of the Univerſe, eſpecially if we do but conſider, that what he wrote, was only to the Senſes of a People who had not yet learnt to make uſe of their Reaſon any other way, but from the Appearance of Things, and upon a Subject too ſublime for vulgar Capacities in any Age, and had only been attempted in the deepeſt Learning of *Egypt*, which, he though well acquainted with, the Generality of them were totally Strangers to.

In the firſt Place it muſt be granted, that the Stars being all of the ſame Nature, are either all immoveable, or all fixed, that is all governed by one and the ſame Principle.

Now to ſuppoſe them all fixed, and diſperſed in an endleſs Diſorder thro' the infinite Expanſe, which has long been the Opinion of many very able Aſtronomers amongſt the Antients, and even now received by too many of the Moderns, implies an Inactivity in thoſe vaſt and principal Bodies, ſo much the Reverſe of what may be expected, and what we daily obſerve through all the reſt of their Attendants, namely, their own reſpective Satellites, that we cannot poſſibly upon any rational Grounds, advance one ſingle Argument to ſupport ſo much as a Conjecture towards it, without betraying the greateſt Simplicity, and next to an Affirmation reduce the whole Frame of Nature, and all corporeal Beings to a wild unmeaning Chance, ariſing from an unnatural Diſcord and Confuſion.

For upon the Principles of Locality and Materiality, you having allowed me the Uſe of my Senſes and Reaſon, as abſolutely neceſſary towards conceiving any Idea of our preſent State, or of Futurity: Upon

* The Moon, Satellites of *Saturn* and *Jupiter*, &c.

these Principles I say, unless our Faculties are useless, if there are no other Bodies or Beings in the Universe than what we see, and are now sensible of, we must now at the Height of this our present State, be as near Perfection as we can reasonably expect, and as such ourselves the supreme Beings of all Beings. To what End then do we form Ideas of a succeeding Life, where a more exalted State cannot be hoped for.

How absurd and impious this is I leave to your own Reason and Reflection: This is the fatal Rook upon which all weak Heads and narrow Minds are lost and split upon, consequently ought to be the most carefully avoided, not only as the Nurse of Atheism, but as the dreadful Father of Despair: " For, say they, these unhappy Wretches, to be always the " same, is inconsistent with a Change; and to be less than what we are, " any where hereafter, is full as difficult to conceive as to be more." Thus, unless we admit of superior Seats and much more glorious Habitations than these we are sensible of, we strike at the very Root of a fair flourishing Tree of Immortality, and must become Authors of our own Despair. I have often wonder'd how thinking Men could possibly fall into so gross an Error, as that of a Spirit's Annihilation; and I should be glad to ask one of those fruitless Students, whether, upon the Evidence of our present Being, it is not much more rational, to hope for a future, than to expect a *Ne plus ultra* upon no Evidence at all. The Affirmative is certainly much more natural to be conceiv'd than the Negative. But if Chance were the Case, and that Chance produced all these regular and wondrous Works, 'tis to be wished at least, that Chance might do the same again; and if not Chance, of course an eternal Direction: But Chance only can effect Disorder, Discord, and Confusion; *ergo*, the visible Harmony and Beauty of the Creation declare for a Direction; and this must of Consequence, from its perfect Nature, proceed from the Wisdom and Power of an eternal Being, *God of Infinity*, the Author of all Ideas: And if this primitive Power produced us his Creatures from nothing, nothing can be wanting to revive our Frames again; and if from something, that something must remain to establish us in a future Life. But to return, how absurd it is to suppose one Part of the Creation regular, and the other irregular, or a visible circulating Order of Things, to be mixed with Disorder, and circumscribing Part of an endless Confusion, is obvious to the weakest Understanding, and consequently we may reasonably expect, that the *Via Lactea*, which is a manifest Circle amongst the Stars, conspicuous to every Eye, will prove at last the Whole to be together a vast and glorious regular Production of Beings, out of the wondrous Will or Fecundity of the eternal and infinite *one* self-sufficient Cause; and that all its Irregularities are only such as naturally arise from our excentric View: To demonstrate which

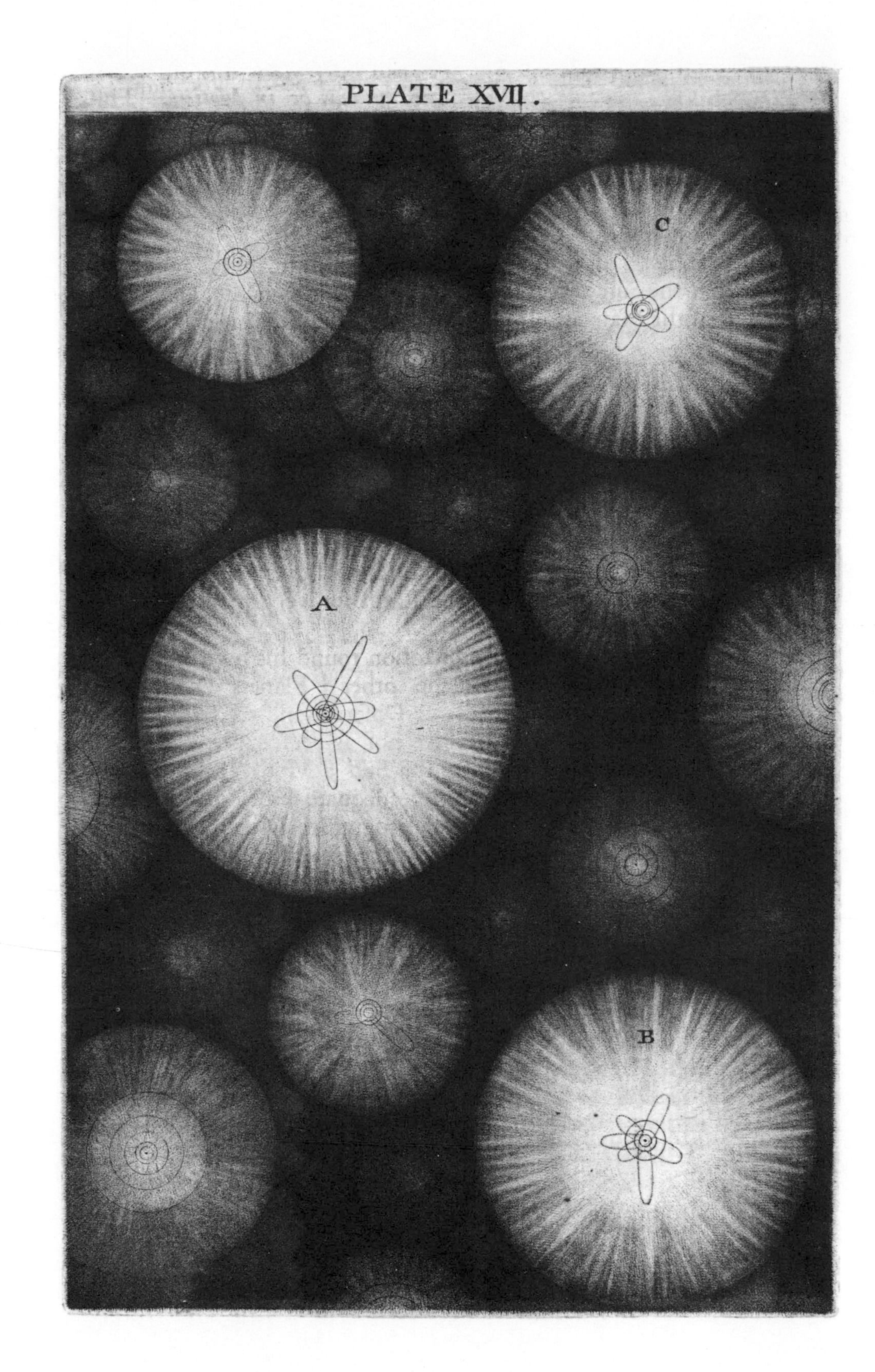

PLATE XVII.

which abſolutely and inconteſtibly, we ſhall only want this one *Poſtulata* to be granted, *viz. That all the Stars are, or may be in Motion:* This, if one may be allowed to judge of the Whole by the Similitude and Government of its Parts, I am perſwaded you will think a very reaſonable Aſſumption; but that you may imbibe a good Opinion of this Aſſumption, and entirely come into this much better to be wiſhed Hypotheſis, I would have you conſult theſe following Arguments.

Firſt, it is allowed, as I have endeavoured to ſhew, by all modern Philoſophers, that the Sun and Stars are all of the ſame or like Nature; conſequently, that the Stars are all Suns, and that the Sun himſelf is a Star.

PLATE XVII.

Repreſents a kind of perſpective View of the viſible Creation, wherein A repreſents the Syſtem of our Sun, B, that ſuppoſed round *Syrius*, and C, the Region about *Rigel*. The reſt is a promiſcuous Diſpoſition of all the Variety of other Syſtems within our finite Viſion, as they are ſuppoſed to be poſited behind one another, in the infinite Space, and round every viſible Star. That round every Star then we may juſtly conjecture a ſimilar Syſtem of Bodies, governed by the ſame Laws and Principles with this our ſolar one, though to us at the Earth for very good Reaſons inviſible *. Secondly,

The Sun is alſo obſerved to have a Motion round his own Axis in about twenty-five Days. Now, ſince all the other † Planets which move in Orbits round him, and are within our Obſervation, are found to have a like Rotation round their Axis, may we not as reaſonably imagine, that that Power which was able to give the Sun a Motion round his Axis, could and would at the ſame time, with adequate Eaſe, give him alſo an orbitular one? and why not, ſince no progreſſive Mutability can either take from, or diſturb the boundleſs Property of an Infinity; and beſides, ſeeing to imagine him at reſt, is to impoſe ſuch an unnatural Stagnation upon the eternal Faculty, quite repugnant to that imparable Power which we ſuppoſe ſtands in need of neither Sleep nor Reſt?

'Tis true, the Sun may be ſaid to be the Governor of all thoſe Bodies round him; but how? no otherwiſe than he himſelf may be governed by a ſuperior Agent, or a ſtill more active Force; and methinks it is not a

* *Anaximines* believed the Stars to be of a fiery Nature; and that there were certain terreſtrial Bodies that are not ſeen by us, carried together round them. *Stob. Ecl. Phyſ.* cap. 25. *Pythagoras* affirmed, that every Star is a World, containing Earth, Air, and Æther.

† *Saturn, Jupiter, Mars, Venus,* the Earth, Moon, and *Mercury*.

H 2 little

little abſurd to ſuppoſe he is not, ſince we have diſcovered by undoubted Obſervations, that the ſame gravitating Power is common to all; and that the Stars themſelves are ſubject to no other Direction than that which moves the whole Machine of Nature.

Thirdly, From many Obſervations of the polar Points, and the Obliquity of the Earth's Equator to the Plane of her ſolar Orbit compared together, the Sun is very juſtly ſuſpected to have changed his ſidereal Situation; and this muſt either ariſe from a Change in the Poſition of the Earth's diurnal Axis, or from a Removal of the Sun himſelf, out of the primitive Plane of the *Orbis Magnus*. I believe you are ſo much of a Mathematician, as to know that if either of theſe Facts be allowed, the Conſequence I want will follow. I ſhall not therefore here enter into any farther Diſpute about it; but I think it will be neceſſary to ſubmit ſome Obſervations to your Conſideration, that may convince you that there is a Motion ſomewhere to be thus diſcovered, and whether in the Sun, or in the Stars, or in both, I leave to your own Determination, but to aſſiſt your Imagination, I refer you to

PLATE XVIII.

The Globe S is here ſuppoſed to repreſent the Sun, having changed its Situation by a local Motion from A to C, and B repreſents the Globe of the Earth in a permanent Poſition, with its principal Points and Circles, reſpecting the primitive Plane A, B, K. Now in Conſequence of the Angle of Variation, A, B, C, it evidently appears that a new ecliptic Plane, will be produced, as C, B, and alſo a Variation in the greateſt Declination of the Sun, North and South from the Line of the *Equator* D, L. Hence, as in this Figure, the Obliquity of the Poles P, N, and G, F, will naturally decreaſe, and is ſhewn in Quantity by the Line of Aberration H, I.

Here follows a Table of the Change obſerved in the Obliquity of the Ecliptic by Aſtronomers of different Ages.

A Table of the Obliquity of the Ecliptic.

Ante Chriſti		°	′
124	ARATO - - - - - - - - - - - - -	24	00
——	HIPARCHUS - - - - - - - - - - -	23	51 $\frac{1}{3}$
127	ERATOSTHENES - - - - - - - - - -	23	51 $\frac{1}{2}$

Anno

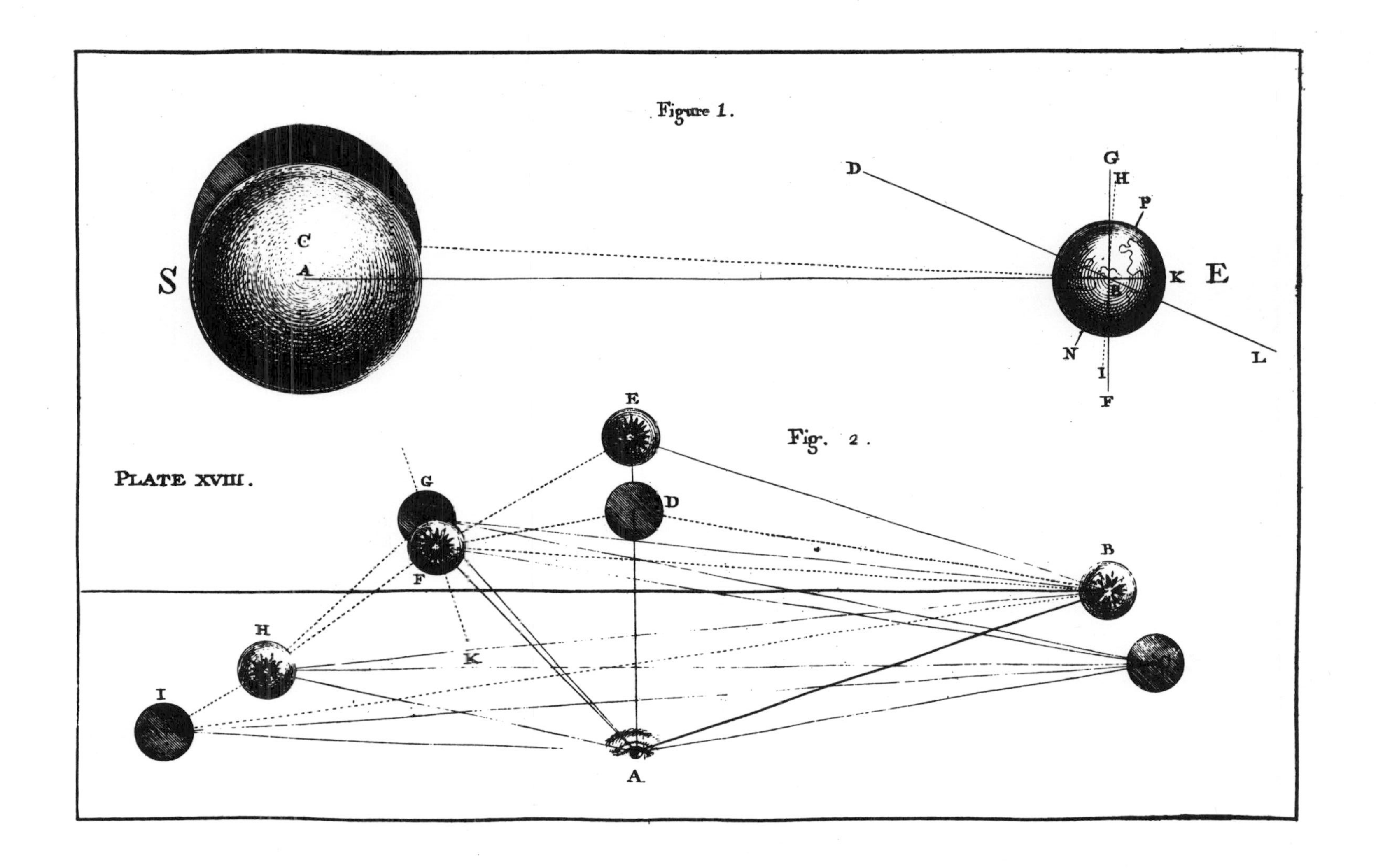
Figure 1.
S
C
A
D
G
H
P
K
E
B
N
I
F
L
Fig. 2.
PLATE XVIII.
E
D
G
F
B
H
K
I
A

Anno Dom.		°	′
140	PTOLOMY - - - - - - - - - - -	23	51 1/3
749	ABATEGNIUS - - - - - - - - - - -	23	35 1/2
1070	AIRAHEL - - - - - - - - - - - -	23	34
1140	ALOMEAN - - - - - - - - - - - -	23	33
1300	PROFATIOGRAD - - - - - - - - -	23	32
1458	PURBACCHIO - - - - - - - - - - -	23	29 1/2
1490	REGIOMONTAUS - - - - - - - - - -	23	30
1500	COPERNICUS - - - - - - - - - - -	23	28 1/2
1592	TYCHO BRAHE - - - - - - -	23	21 1/2
1656	CASSINI - - - - - - -	23	29 1/2

Now ſure, if we conſider this continual Decreaſe of the Sun's Declination, which can proceed from no other Cauſe than that of his having moved out of the primitive Plane; we need make no great Difficulty thus far, to think our Conjectures not irrational.

The following is a Citation from Dr. *Edmund Hally*, Aſtronomer-Royal. See *Philoſophical Tranſactions*, N°. 355. p. 736.

" But while I was upon this Enquiry (*of the Obliquity of the Ecliptic*) I was ſurprized to find the Latitudes of three of the principal Stars in the Heavens, directly to contradict the ſuppoſed greater Obliquity of the Ecliptic, which ſeems confirmed by the Latitudes of moſt of the reſt; they being ſet down in the old Catalogues, as if the Plane of the Earth's Orbit had changed its Situation amongſt the fixed Stars, about 20′ ſince the Time of *Hipparchus*, particularly all the Stars in *Gemini* are put down, thoſe to the Northward of the Ecliptic, with ſo much leſs Latitude than we find, and thoſe to the Southward, with ſo much more ſoutherly Latitude; and yet the three Stars *Palilicium*, *Sirius*, and *Arcturus*, do contradict this Rule: For by it, *Palilicium*, being in the Days of *Hipparchus*, in about 10 gr. of *Taurus*, ought to be about 15′ more ſoutherly than at preſent, and *Sirius* being then in about 15 gr. of *Gemini*, ought to be 20′ more ſoutherly than now; yet *Ptolomy* places the firſt 20′, and the other 22′ more northerly in Latitude than we now find them: Nor are theſe the Errors of Tranſcribers, but are proved to be right by the Declination of them ſet down by *Ptolomy*, as obſerved by *Timocharis*, *Hipparchus*, and himſelf; which ſhew, that theſe Latitudes are the ſame as thoſe Authors intended. As to *Arcturus*, he is too near the Equinoctial Colour, to argue from him concerning the Change of the Obliquity of the Ecliptic; but *Ptolomy* gives him 33′ more North Latitude than

than he is now ſound to have; and that greater Latitude is likewiſe confirmed by the Declinations delivered by the aboveſaid Obſervations: So then theſe three Stars are found to be above half a Degree more ſoutherly at this Time than the Antients reckoned them. When, on the contrary, at the ſame time, the bright Shoulder of *Orion*, has, in *Ptolomy* almoſt a Degree more ſoutherly Latitude than at preſent, what ſhall we ſay then? It is ſcarce to be believed, that the Antients could be deceived in ſo plain a Matter, three Obſervers confirming each other. Again, theſe Stars being the moſt conſpicuous in Heaven, are in all Probability the neareſt to the Earth; and if they have any particular Motion of their own, it is moſt likely to be perceived in them, which in ſo long a Time as eighteen hundred Years, may ſhew itſelf by the Alteration of their Places, though it be intirely imperceptible in the Space of one ſingle Century of Years: Yet, as to *Syrius*, it may be obſerved, that *Tycho Brahe* makes him 2 Min. more northerly than we now find him; whereas he ought to be above as much more ſoutherly from his Ecliptic (whoſe Obliquity he makes 2′ ½ greater than we eſteem it at at preſent) differing in the Whole 4′ ½.

One Half of this Difference may perhaps be excuſed, if Refraction were not allowed in this Caſe by *Tycho*; yet 2 Min. in ſuch a Star as *Syrius*, is ſomewhat too much for him to be miſtaken in.

But a more evident Proof of this Change is drawn from the Obſervation of the Application of the Moon to *Pal.licium*, *An. Chriſ.* 509. *Mar.* 11. when, in the Beginning of the Night, the Moon was ſeen to follow that Star very near, and ſeemed to have eclipſed it, ἐπέβαλλε γὰρ ὁ ἀστηρ τῳ πᾶρα την διχοτομίαν μέρει τῆς κυ'ρτυς περιφειας τοῦ πεφωτισμένου μερους, *i. e. Stella appoſita erat parti per quam biſecabatur limbus Lunæ illuminatus*, as *Bullialdus*, to whom we are beholden for this ancient Obſervation, has tranſlated it. Now, from the undoubted Principles of Aſtronomy, this could never be true at *Athens*, or near it, unleſs the Latitude of *Palilicium* were much leſs than we at this Time find it *.

The Motion of *Arcturus* ſeems further confirmed, from the Obſervations of *Tycho Hevelius* and *Flamſtead*; for *Hevelius* ſets down the Diſtance of that Star from *Lyra* 4′ greater than *Tycho* had obſerved it ſeventy-two Years before him, and *Flamſtead* twenty-two Years after meaſured

* Vide *Bulialdi Aſtr. Philolaica*, p. 172.

† Theſe are the neareſt and greateſt of the fixed Stars, the Motion of the others not having been obſerved, or being at too great a Diſtance, are either imperceptible, or have not been taken notice of.

the

the Diſtance betwixt the ſame two Stars, ſtill 3′ greater than *Hevelius* found it ; ſo that if *Lyra* had ſtood ſtill all that while, there was an Appearance of *Arcturus*'s having gone 7′ out of his Place in the Space of an hundred Years. See Dr. *Long*'s Aſtronomy, p. 274.

It is further to be obſerved, in Confirmation of the Motion of one of theſe Stars, that *Flamſtead* found the Diſtance of *Arcturus*, from the Head of *Hercules* 3′ greater than it is ſet down by the Prince of *Heſſe* ; and that his Diſtance from the *Lion's Tail* was a little decreaſed with $5'\frac{1}{2}$ leſs Latitude than *Tycho* had obſerved. Hence, to make theſe Obſervations agree, one or both of them muſt have moved together equal to 7′. This Change of Place, which is quite contrary to all known Cauſes proceeding from the Earth, muſt therefore be occaſioned either by the Motion of the Sun, or by a particular Motion of their own ; but if, amongſt themſelves, they muſt all move, and if all be in Motion, the Sun muſt alſo move.

If theſe Obſervations, delivered down to us by very able Aſtronomers, be either true or near it, as great Allowances have been made for the Ignorance of the Ages in which they were taken, and the Inaccuracy of the Inſtruments, we may naturally conclude, that theſe Stars muſt have a Motion ; and if they move, as has been before obſerved, the Sun muſt alſo ; hence he cannot now be in the original Plane of the Earth's annual Direction, or at leaſt in the ſame identical Place he was at firſt poſſeſſed of : And if ſo, the Stars muſt alſo have the like Motion, though in different Directions, and all may thus be governed by the ſame impulſive Power.

To illuſtrate this primitive Motion of the Stars, and at the ſame time to ſhow that the Variety which appears in the Quantity of Motion can be no Objection to it,

See PLATE XVIII. *Fig.* 2.

Where A repreſents the Eye of an Obſerver, and B, E, F, H, various Syſtems, moving in different Directions thro' the mundane Space; it is evident that the Sphere B, having moved from C, and that of E, not having appeared to move at all, there muſt be a ſenſible Change in the new Poſition of theſe two Syſtems to one another, and ſo of the reſt ; and tho' the apparent Motion of H, be much more than that of F, from the Point A, yet from C, they will appear leſs different, and from B, they will appear nearly equal. And farther, as the Direction from H, is in the Line I, H, and that of F, in the Line K, G, thoſe two Syſtems will appear to approximate, and the Magnitude of the Star in the firſt will be increaſed, creafed, and in the latter diminiſhed. Thus, many of the Stars in the oldeſt Catalogues, which were ſaid to be of the ſecond Magnitude, are now become of the firſt, and ſeveral of the firſt are now judged to be of the ſecond, &c.

But as this apparent Motion of the Stars at the Earth, muſt, from its Nature, be very ſmall, ſo as ſcarce to be diſcovered in ſome of them in leſs than an Age, with any Inſtrument by the niceſt Obſerver, I judge it will be extremely proper in this Place to propoſe ſome Method, by which, in proceſs of Time, the Truth of the Theory may be aſcertained. The Way I think moſt likely to ſucceed is this.

PLATE XIX.

Is a Plan of the principal Stars that form the PLEIADES, correctly taken by a Combination of Triangles, as in the Figure, from whence it will naturally follow, all the whole Form being comprehended in much leſs than one Degree. That the moſt minute local Motion in any one of thoſe Stars in a very few Years, will be made ſenſible to an Eye at the Earth. For Inſtance, if any of the Stars that form the Letter A, or T, within the Term of ten or twenty Years, be found in the leaſt to deviate from the Lines of their preſent Poſition and Direction, it will be evident beyond a Contradiction, that they have a Motion amongſt themſelves, and ſince at ſuch a Diſtance they cannot poſſibly be affected by the Earth, it muſt be a Motion of their own; and thus if any one can be proved, to change its Situation, with regard to the reſt, we can have no new Difficulty in concluding that they all may do the ſame.

Thus if any of the regular Triangles M B Z, Z P H, A Z M, T A Γ, or Π O I, &c. in due Time be carefully noted, we may venture to ſay with great Safety, that the thouſandth Part of a Degree will be plainly diſcovered.

PLATE XX.

Is a true Plan and Combination of the principal Stars that form the PERSEDES, in which other Obſervations may be made in a different Part of the Heavens, and perhaps with an Opportunity of being ſtill more exact, the Areas of theſe Triangles, particularly that of Θ I K, and thoſe of ρ and δ, being much leſs than the former, where the leaſt Alteration poſſible muſt render them ſenſibly diſtorted. But here it muſt be conſidered, that the real Motion of the Stars, as well as their apparent, may be, and in all

PLATE. XIX.

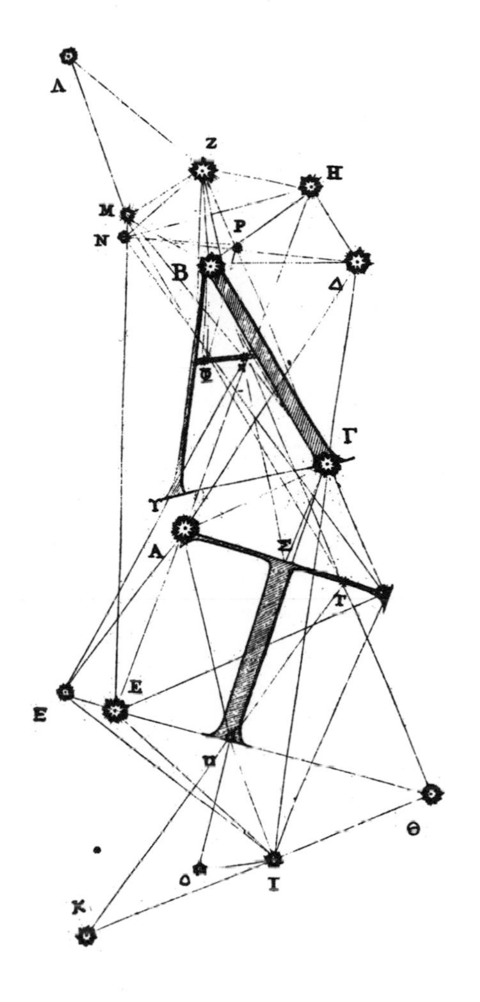

PLATE. XX.

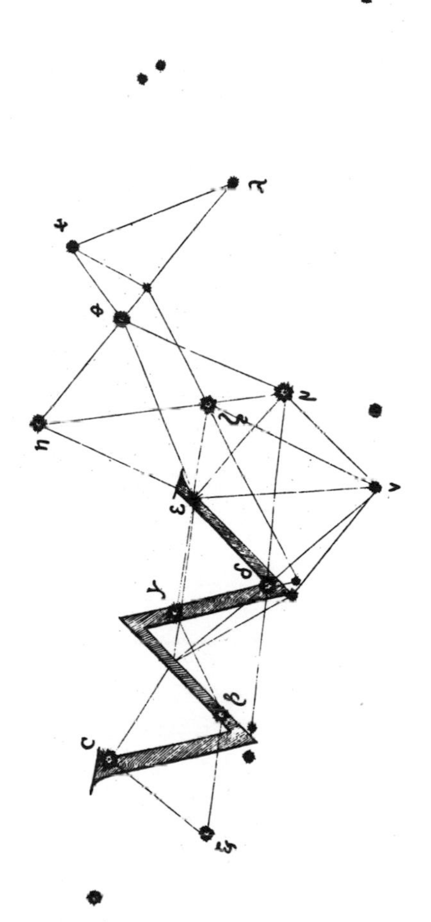

all Likelihood, is extreamly ſlow, for the moſt minute, viſible, local Motion, will anſwer all the Purpoſes we know in Nature, and the greateſt ſeems to be that of the projectile, or centrifugal Force, which not only preſerves them in their Orbits, but prevents them from ruſhing all together, by the common univerſal Law of Gravity, which otherwiſe, as a finite Diſtribution of either regular or irregular Bodies, they muſt at length do by Neceſſity.

I muſt now inform you, that the above Obſervations were compleated in the AUTUMN SEASON, 1747, and were taken by myſelf; the Letters A, T, in *Plate* XIX, and the W in the XXth, as you may ſee, having a very near Reſemblance, or Similitude, to the Order theſe Stars are found to be in, together with the *Greek* Alphabet, I judged neceſſary, by way of *Aſteriſm* and *Nomenclatura*, in caſe ſuch ſhould be wanted, as *Data* in future Diſcoveries.

I come now to the principal Point in Queſtion, which is to find a regular Diſpoſition of the Stars amongſt themſelves, which will naturally ſolve both their general and particular Phænomena, eſpecially the *Nebulæ* and *Milky Way*.

I am now, &c.

I LETTER

LETTER THE SEVENTH

The Hypothesis, or Theory, fully explained and demonstrated, proving the sidereal Creation to be finite.

SIR,

I KNOW you are an Enemy to all Sorts of Schemes where they are not absolutely necessary, and may possibly be avoided; and for that Reason I have purposely omitted many geometrical Figures, and other Representations in this Work, which might have been inserted and in some Places, especially here I might have introduced Diagrams, perhaps more explicit than Words; but as you have frequently observed, they are only of Use to the few Learned, and contribute more to the taking away the little Ideas and Knowledge the more ignorant Many may be endued with, by a prejudicial Impression of imperfect Images, rather than the adding any new Light to their Understanding, I have purposely avoided, as much as possible, both here and every where, all such complex Diagrams as might be in Danger of betraying any the least such conscious Diffidence in you, arising from the Want of a proper *Precognita* in the Sciences.

This Imperfection, much to be lamented, as greatly to the Disadvantage of all mathematical Reasoning, I would willingly always prevent, in my Readers, and to chuse in my Friend; I shall therefore content myself with referring you to a few orbicular Figures, concave and convex, as may best suggest to your Fancy the simplest Way, a just Idea of the Hypothesis I have fram'd, and naturally enough I hope, render my Theory so intelligible, as to help you sufficiently to conceive the Solution aimed at, of the important Problem I have attempted.

As I have said before, we cannot long observe the beauteous Parts of the visible Creation, not only those of this World on which we live, but also the Myriads of bright Bodies round us, with any Attention, without being convinced, that a Power supreme, and of a Nature unknown to us, presides in, and governs it.

The

The Courſe and Frame of this vaſt Bulk, diſplay
A Reaſon and fix'd Law, which all obey.

SHER. MANILIUS.

And notwithſtanding the many wonderful Productions of Nature in this our known Habitation, yet the Earth, when compared with other Bodies of our own Syſtem, ſeems far from being the moſt conſiderable in it; and it appears not only very poſſible, but highly probable, from what has been ſaid, and from what we can farther demonſtrate, that there is as great a Multiplicity of Worlds, variouſly diſperſed in different Parts of the Univerſe, as there are variegated Objects in this we live upon. Now, as we have no Reaſon to ſuppoſe, that the Nature of our Sun is different from that of the reſt of the Stars; and ſince we can no way prove him ſuperior even to the leaſt of thoſe ſurpriſing Bodies, how can we, with any Shew of Reaſon, imagine him to be the general Center of the whole, *i. e.* of the viſible Creation, and ſeated in the Center of the mundane Space? This, in my humble Opinion, is too weak even for Conjecture, their apparent Diſtribution, and * irregular Order argue ſo much againſt it.

The Earth indeed has long poſſeſſed the chief Seat of our Syſtem, and peaceably reigned there, as in the Center of the Univerſe for many Ages paſt; but it was human Ignorance, and not divine Wiſdom, that placed it there; ſome few indeed from the Beginning have diſputed its Right to it, as judging it no way worthy of ſuch high Eminence. Time at length has diſcovered the Truth to every body, and now it is juſtly diſplaced by the united Conſent of all its Inhabitants, and inſtead of being thought the moſt majeſtick of all Nature's lower Works, now rather diſgraces the Creation, ſo much it is reduced in its preſent State from what it had Reaſon to expect in the former.

Now it is no longer the only terreſtrial Globe in the Univerſe, but is proved to be one of the leaſt Planets of the ſolar Syſtem, and ſurprizingly inferior to ſome of its Fellow Worlds. The Sun, or rather the Syſtem, has almoſt as long uſurped the Center of Infinity, with as little Pretence to ſuch Pre-heminence; but now, Thanks to the Sciences, the Scene begins to open to us on all Sides, and Truths ſcarce to have been dreamt of, before Perſons of Obſervation had proved them poſſible, invades our Senſes

* See the Zodaical Conſtellations, you'll find that in ſome Signs there are ſeveral Stars of the firſt, ſecond, and third Magnitude, and in many others none of theſe at all.

with a Subject too deep for the human Underſtanding, and where our very Reaſon is loſt in infinite Wonders. How ought this to humble every Mind ſuſceptible of Reaſon!

In this Place, I believe, you will pardon a Digreſſion; which, in Anſwer to Part of your laſt Letter, I judge will not be very impertinent, tho' perhaps juſt here I cannot ſo well juſtify it.

Your late Converſation with our Friend Mr. * * *, I am perſwaded, muſt have been very entertaining; but I cannot help thinking his Reflections upon the Wonders of Nature and the Wiſdom of Providence, though I muſt allow them all to be very juſt and curious, inſtead of elevating the Mind to the Pitch he would have it, rather as conſidered above, depreſs it below the proper, nay I might ſay neceſſary, Standard of human Ideas.

This, probably, you'll ſay is an odd Turn, and may want ſome Explanation, ſince every Object in the Chain of Nature, muſt of Force be granted, a Subject worthy of our Speculations, being all together made, as in the Maximum of Wiſdom: But what I mean is this, ſince nothing is more natural for Beings in every State in ſearch after their own Advantages, and the Enlargement of their Ideas to look upward, ſure it may be preſumed, that Time may be miſpent, if not loſt in inſpecting too narrowly Things ſo little benefical in States below us; as Mr. *Pope* ſays,

> Why has not Man a microſcopic Eye?
> For this plain Reaſon, Man is not a Fly.
> Say what the Uſe, were finer Opticks given,
> To inſpect a Mite, not comprehend the Heav'n.
>
> *Eſſay on Man.*

Amuſement alone can never be ſuppoſed to be the ſole End of human Life, where even true Happineſs is a Thing we rather taſte than enjoy. The Mind we find capable of much more rational Pleaſure than can poſſibly fall within the Reach of human Power, either to promiſe or procure it; but then this very Defect in our preſent State of Exiſtence affords us no leſs than a moral Aſſurance, that ſome where in a future, we may, if we pleaſe, be entitled to the very *Plenum* of all Enjoyments.

The peculiar Buſineſs then of the human Mind naturally precedes its Amuſements, as evidently ordained to ſoar above all the inferior Beings of this World; and however our Natures may, thro' Indolence, or thro' Ignorance, degenerate, that of the Man can never be ſuppoſed to ſink into the Mole.

The propereſt Way then ſure for Men to preſerve their Pre-heminence over the Brute Creation, is to make uſe of that Reaſon and Reflection, which

which ſo manifeſtly diſtinguiſhes their natural Superiority. A right Application of which, muſt of courſe then direct us to a forward, rather than a backward Search in the vaſt viſible Chain of our Exiſtence, which clearly connects all Beings and States as under the Direction of one ſupreme Agent.

This is all I would have underſtood by the foregoing Poſition, which, in one Word, implies no more than that the ſublime Philoſophy ought in all Reaſon to be preferred to the Minute; but I hope you will not infer from this my ſeeming Partiality for the celeſtial Sciences, that I mean to inſinuate, that the Study of terreſtrial Phyſicks is not a rational Amuſement.

Mr. ***, you ſay, ſeems to lament the Taſte of Mankind in general much in the ſame Degree as you do his I readily grant you; a Man who can talk ſo well upon an Ant, might make a more entertaining Diſcourſe upon the Eagle; but I beg his Pardon, and though we are all too ready, and moſt apt to condemn all ſuch Pleaſures as vain or trifling, which we have no Share in, or Taſte for ourſelves; yet I don't think it follows, that thoſe ingenious Labours of his are uſeleſs. The Pleaſures ariſing from natural Philoſophy are all undoubtedly great ones, whether we conſider Nature in her higheſt, or in her loweſt Capacity; the Beauties of the Creation are every Day varied to us below, as much they are every Night above, and in both Caſes, through every Object, the Creator ſhines ſo manifeſt, that we may juſtly conſider him every where ſmiling full in the Face of all his Creatures, commanding as it were an awful Reverence, and Reſpect, due not only to his Omnipotency, but alſo to his infinite Goodneſs and endleſs Indulgencies. This is the only Return our Gratitude can make for all thoſe Bleſſings he daily beſtows upon us, and to this great Author of her Laws, Nature herſelf cries aloud through Myriads of various Objects, and after her own expreſſive and peculiar Manner, ſeems to command us with an attractive Grace, to obſerve her Sovereign, and admire his Wiſdom. The Majeſty, Power, and Dominion of God is beſt diſplayed in the external Direction of Things, his Wiſdom and viſible Agency in the internal: Hence, by proper Objects, ſelected from both, attended with juſt Reflections, we may certainly raiſe our Ideas almoſt to the Pitch of Immortals; but how far the human Imagination may poſſibly go, or how much Minds like ours may be improved, is a Queſtion not eaſily determined; but as natural Knowledge evidently increaſes daily, and aſtronomical Enquiries are the moſt capable of opening our Minds, and enlarging our Conception, of conſequence they muſt be moſt worthy our Attention of all other Studies. But of this I have ſaid enough, and think it is now more than Time to attempt the remaining Part of my Theory.

When

When we reflect upon the various Aspects, and perpetual Changes of the Planets, both with regard to their * heliocentric and geocentric Motion, we may readily imagine, that nothing but a like eccentric Position of the Stars could any way produce such an apparently promiscuous Difference in such otherwise regular Bodies. And that in like manner, as the Planets would, if viewed from the Sun, there may be one Place in the Universe to which their Order and primary Motions must appear most regular and most beautiful. Such a Point, I may presume, is not unnatural to be supposed, altho' hitherto we have not been able to produce any absolute Proof of it. See *Plate* XXV.

This is the great Order of Nature, which I shall now endeavour to prove, and thereby solve the Phænomena of the *Via Lactea*; and in order thereto, I want nothing to be granted but what may easily be allowed, namely, that the *Milky Way* is formed of an infinite Number of small Stars.

Let us imagine a vast infinite Gulph, or Medium, every Way extended like a Plane, and inclosed between two Surfaces, nearly even on both Sides, but of such a Depth or Thickness as to occupy a Space equal to the double Radius, or Diameter of the visible Creation, that is to take in one of the smallest Stars each Way, from the middle Station, perpendicular to the Plane's Direction, and, as near as possible, according to our Idea of their true Distance.

But to bring this Image a little lower, and as near as possible level to every Capacity, I mean such as cannot conceive this kind of continued Zodiack, let us suppose the whole Frame of Nature in the Form of an artificial Horizon of a Globe, I don't mean to affirm that it really is so in Fact, but only state the Question thus, to help your Imagination to conceive more aptly what I would explain *. *Plate* XXIII. will then represent a just Section of it. Now in this Space let us imagine all the Stars scattered promiscuously, but at such an adjusted Distance from one another, as to fill up the whole Medium with a kind of regular Irregularity of Objects. And next let us consider what the Consequence would be to an Eye situated near the Center Point, or any where about the middle Plane, as at the Point A. Is it not, think you, very evident, that the Stars would there appear promiscuously dispersed on each Side, and more and more inclining to Disorder, as the Observer would advance his Station towards either Surface, and nearer to B or C, but in the Direction of the general Plane towards H or D, by the continual Approximation of the visual Rays, crowding together as at H, betwixt the Limits D and G, they must in-

* Not to mention their several Conjunctions and Apulces to fixed Stars, *&c.* see the State of the Heavens in 1662, *December* the first, when all the known Planets were in one Sign of the Zodiac, *viz. Sagittarius.*

fallibly

PLATE. XXI.

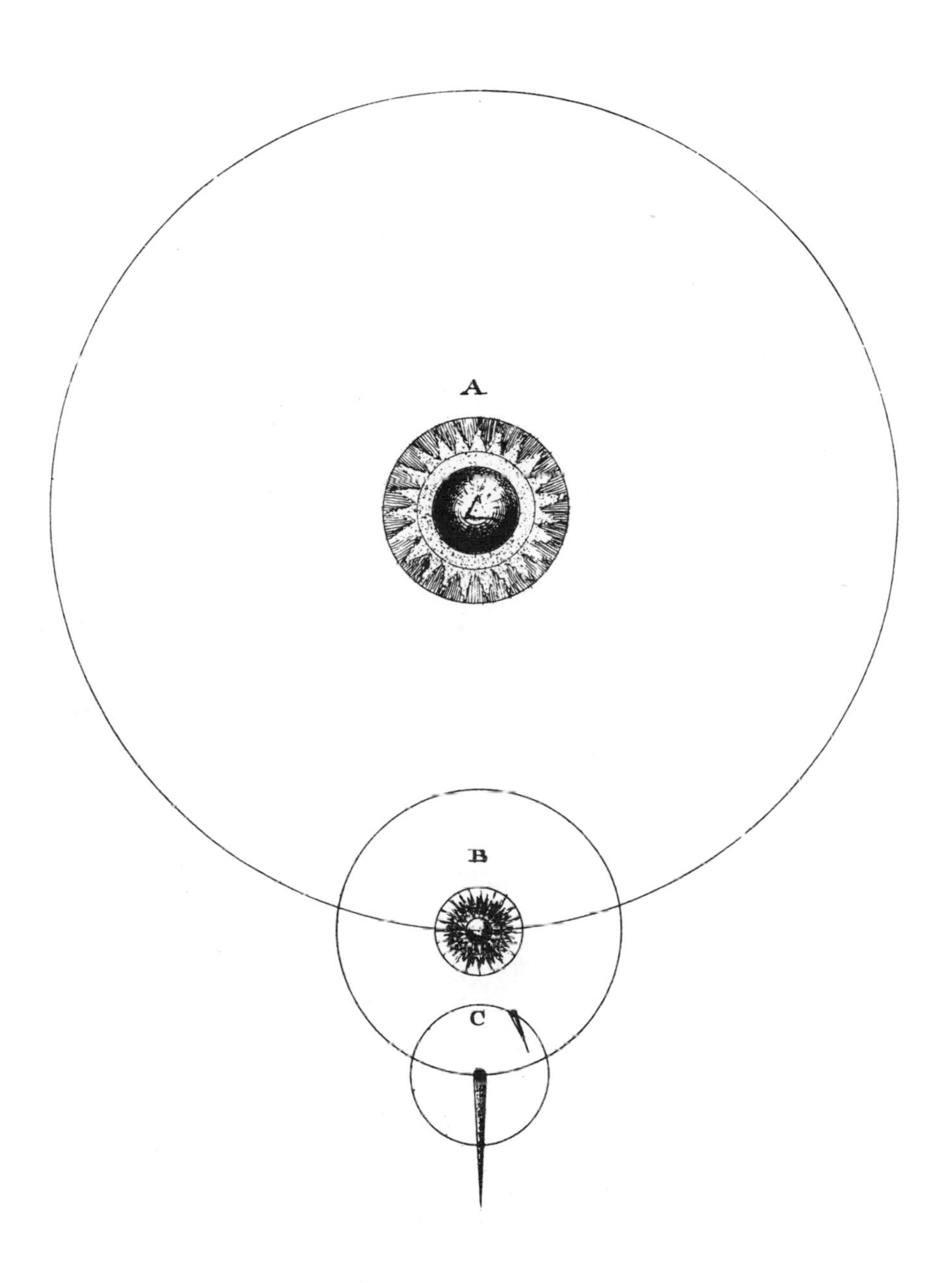

PLATE XXII.

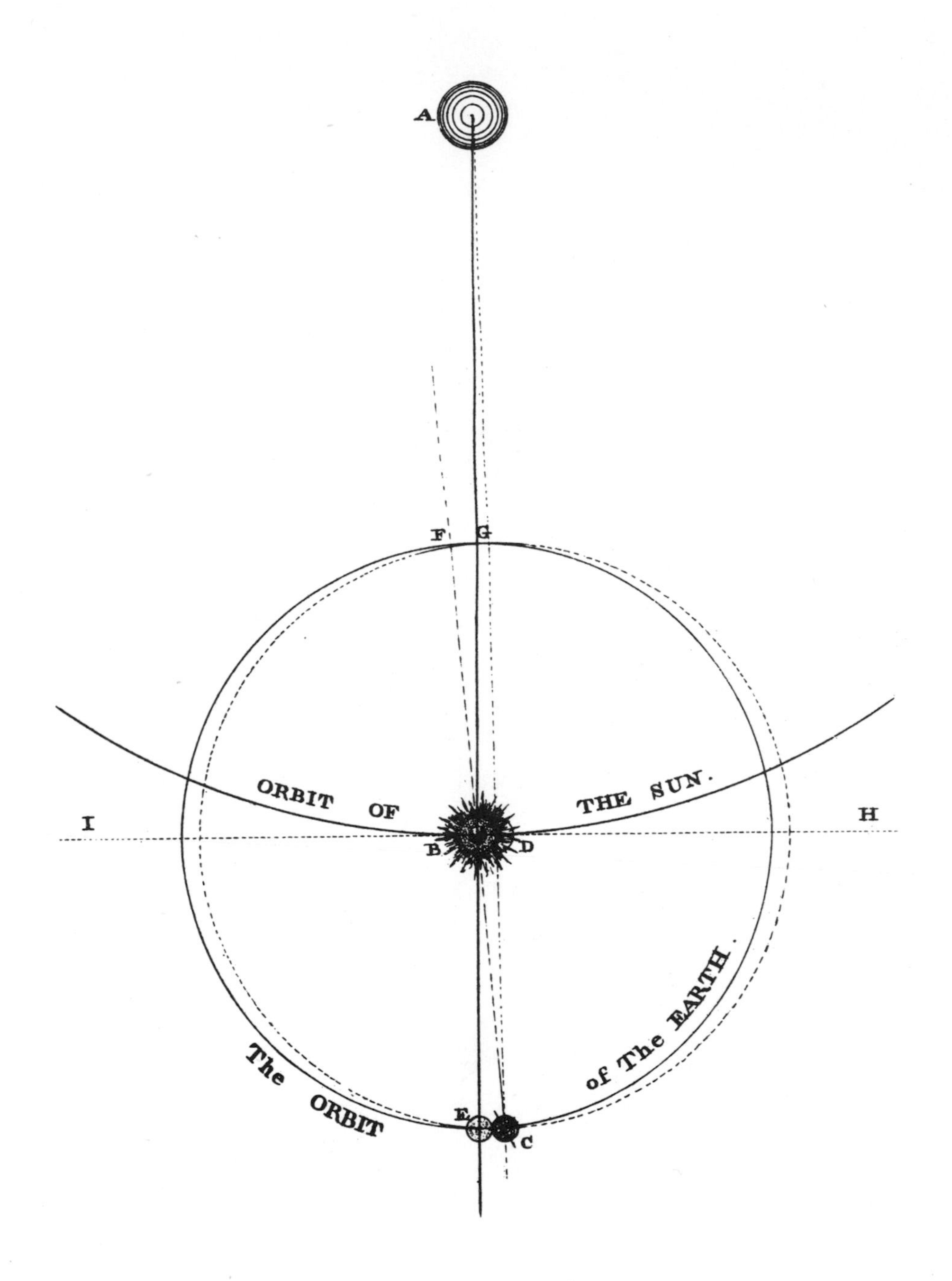

fallibly terminate in the utmoſt Confuſion. If your Opticks fails you before you arrive at theſe external Regions, only imagine how infinitely greater the Number of Stars would be in thoſe remote Parts, ariſing thus from their continual crowding behind one another, as all other Objects do towards the Horizon Point of their Perſpective, which ends but with Infinity: Thus, all their Rays at laſt ſo near uniting, muſt meeting in the Eye appear, as almoſt, in Contact, and form a perfect Zone of Light; this I take to be the real Caſe, and the true Nature of our *Milky Way*, and all the Irregularity we obſerve in it at the Earth, I judge to be intirely owing to our Sun's Poſition in this great Firmament, and may eaſily be ſolved by his Excentricity, and the Diverſity of Motion that may naturally be conceived amongſt the Stars themſelves, which may here and there, in different Parts of the Heavens, occaſion a cloudy Knot of Stars, as perhaps at E.

But now to apply this Hypotheſis to our preſent Purpoſe, and reconcile it to our Ideas of a circular Creation, and the known Laws of orbicular Motion, ſo as to make the Beauty and Harmony of the Whole conſiſtent with the viſible Order of its Parts, our Reaſon muſt now have recourſe to the Analogy of Things. It being once agreed, that the Stars are in Motion, which, as I have endeavoured in my laſt Letter to ſhew is not far from an undeniable Truth, we muſt next conſider in what Manner they move. Firſt then, to ſuppoſe them to move in right Lines, you know is contrary to all the Laws and Principles we at preſent know of; and ſince there are but two Ways that they can poſſibly move in any natural Order, that is, either in right Lines, or in Curves, this being one, it muſt of courſe be the other, *i.e.* in an Orbit; and conſequently, were we able to view them from their middle Poſition, as from the Eye ſeated in the Center of *Plate* XXV. we might expect to find them ſeparately moving in all manner of Directions round a general Center, ſuch as is there repreſented. It only now remains to ſhew how a Number of Stars, ſo diſpoſed in a circular Manner round any given Center, may ſolve the Phænomena before us. There are but two Ways poſſible to be propoſed by which it can be done, and one of which I think is highly probable; but which of the two will meet your Approbation, I ſhall not venture to determine, only here incloſed I intend to ſend you both. The firſt is in the Manner I have above deſcribed, *i.e.* all moving the ſame Way, and not much deviating from the ſame Plane, as the Planets in their heliocentric Motion do round the ſolar Body. In this Caſe the primary, ſecondary, and tertiary conſtituent Orbits, *&c.* framing the Hypotheſes, are repreſented in *Plate* XXII, and the Conſequence of ſuch a Theory ariſing from ſuch an univerſal Law of Motion.

tion in *Plate* XXIII. where B, D denotes the local Motion of the Sun in the true *Orbis Magnus*, and E, C that of the Earth in her proper secondary Orbit, which of course is supposed, as is shewn in the Figure to change its sidereal Positions, in the same Manner as the Moon does round the Earth, and consequently will occasion a kind of Procession, or annual Variation in the Place of the Sun, not unlike that of the Equinoxes, or Motion of all the Stars together, from West to East round the Ecliptic Poles, and probably may in some Degree be the Occasion of it. This Angle is represented, but much magnified, by the Lines F, C, G, and the Unnaturalness, or Absurdity of a right Line Motion of the Sun by the Line I, H.

The second Method of solving this Phænomena, is by a spherical Order of the Stars, all moving with different Direction round one common Center, as the Planets and Comets together do round the Sun, but in a kind of Shell, or concave Orb. The former is easily conceived, from what has been already said, and the latter is as easy to be understood, if you have any Idea of the Segment of a Globe, which the adjacent Figures, will, I hope, assist you to. The Doctrine of these Motions will perhaps be made very obvious to you, by inspecting the following Plates.

PLATE XXIV.

Is a Representation of the Convexity, if I may call it so, of the intire Creation, as a universal Coalition of all the Stars consphered round one general Center, and as all governed by one and the same Law.

PLATE XXV.

Is a centeral Section of the same, with the Eye of Providence seated in the Center, as in the virtual Agent of Creation.

PLATE XXVI.

Represents a Creation of a double Construction, where a superior Order of Bodies C, may be imagined to be circumscribed by the former one A, as possessing a more eminent Seat, and nearer the supream Presence, and consequently of a more perfect Nature. Lastly,

PLATE XXVII.

Represents such a Section, and Segments of the same, as I hope will give you a perfect Idea of what I mean by such a Theory.

Fig. 1. is a corresponding Section of the Part at A, in *Fig.* 2. whose versed Sine is equal to half the Thickness of the starry Vortice A C, or B A. Now I say, by supposing the Thickness of this Shell, 1. you may imagine the middle Semi-Chord A D, or A E, to be nearly 6; and consequently,

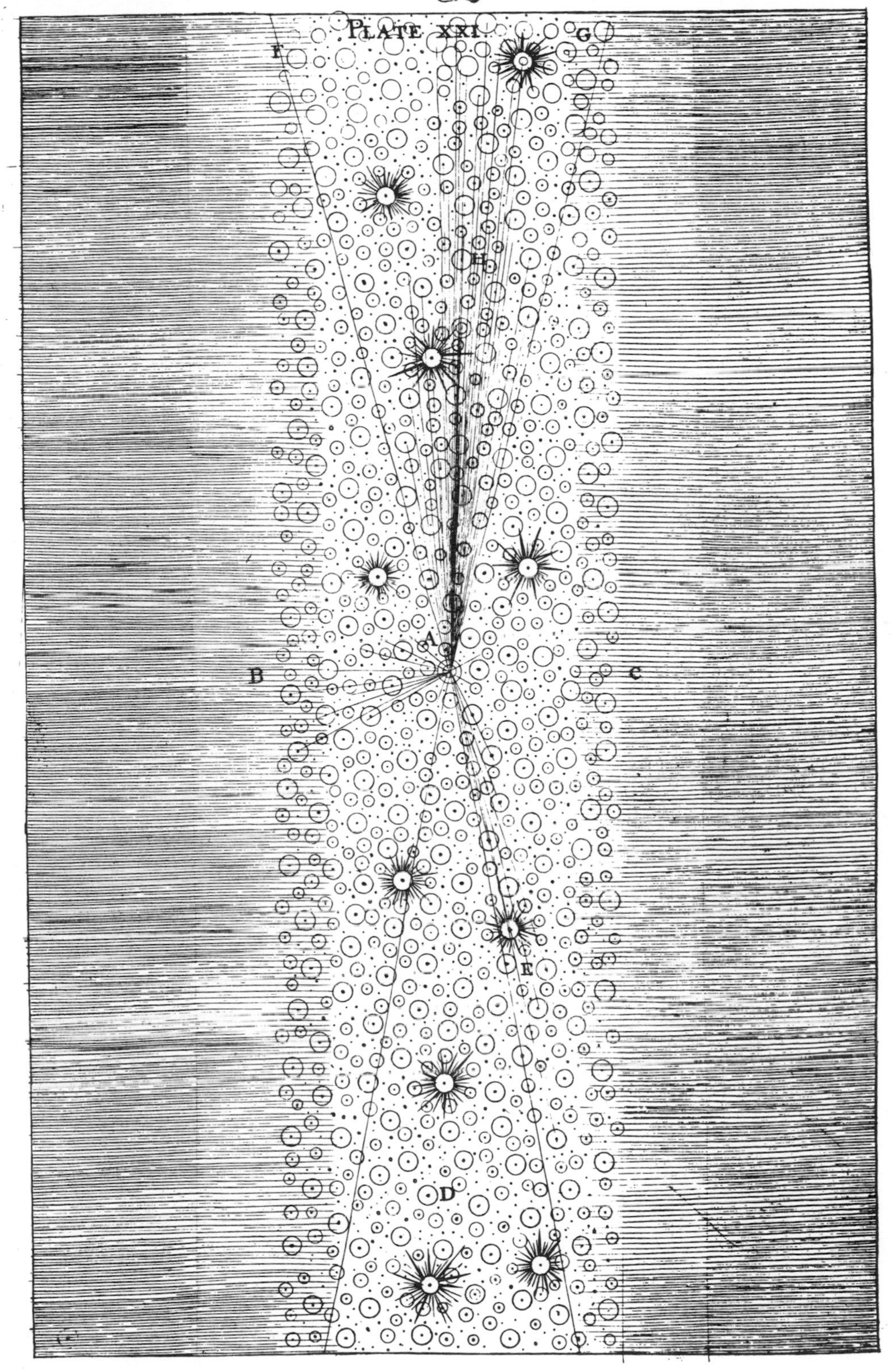
PLATE XXI.
F
G
H
A
B
C
E
D

PLATE XXIV.

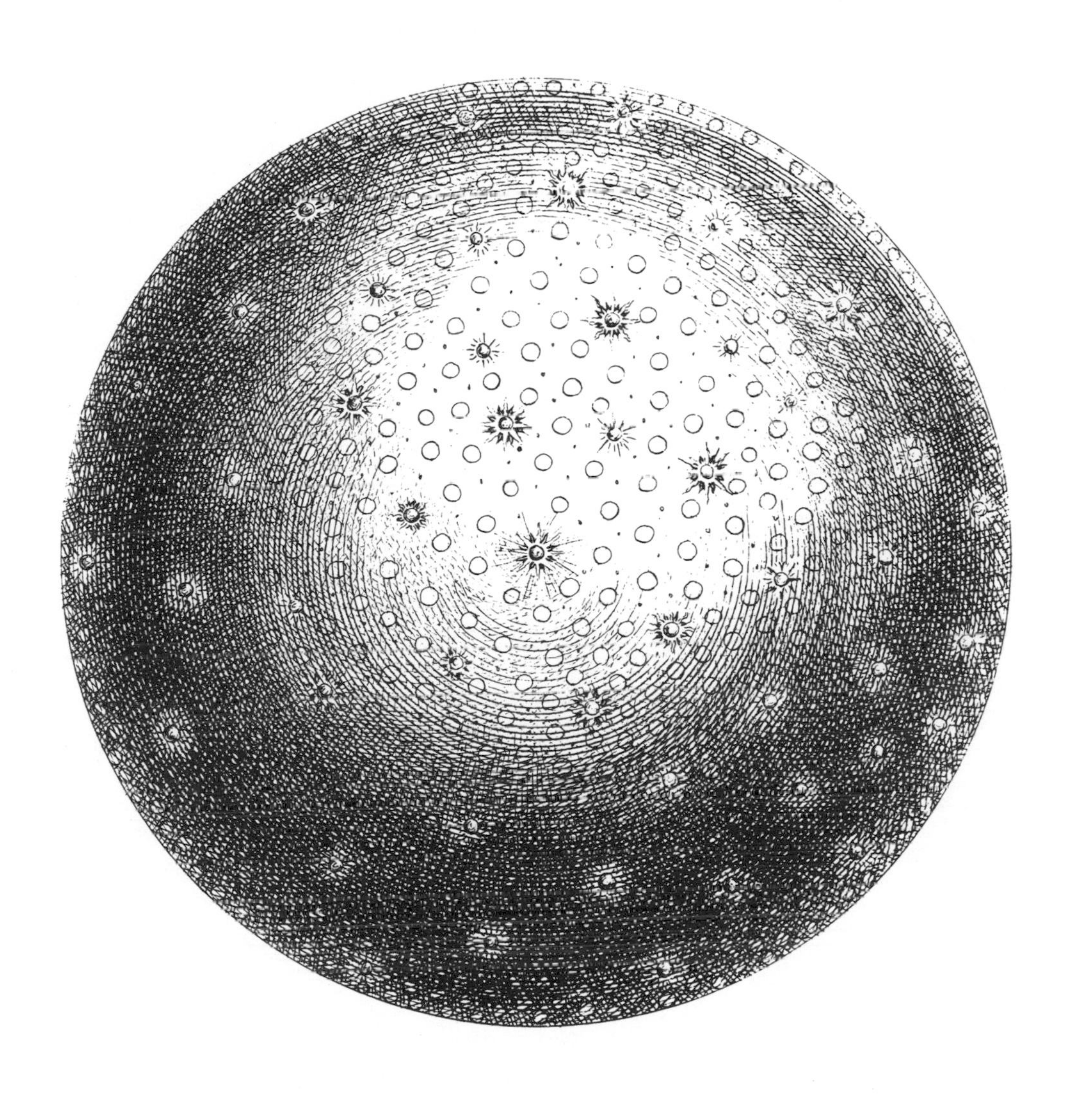

PLATE XXV.

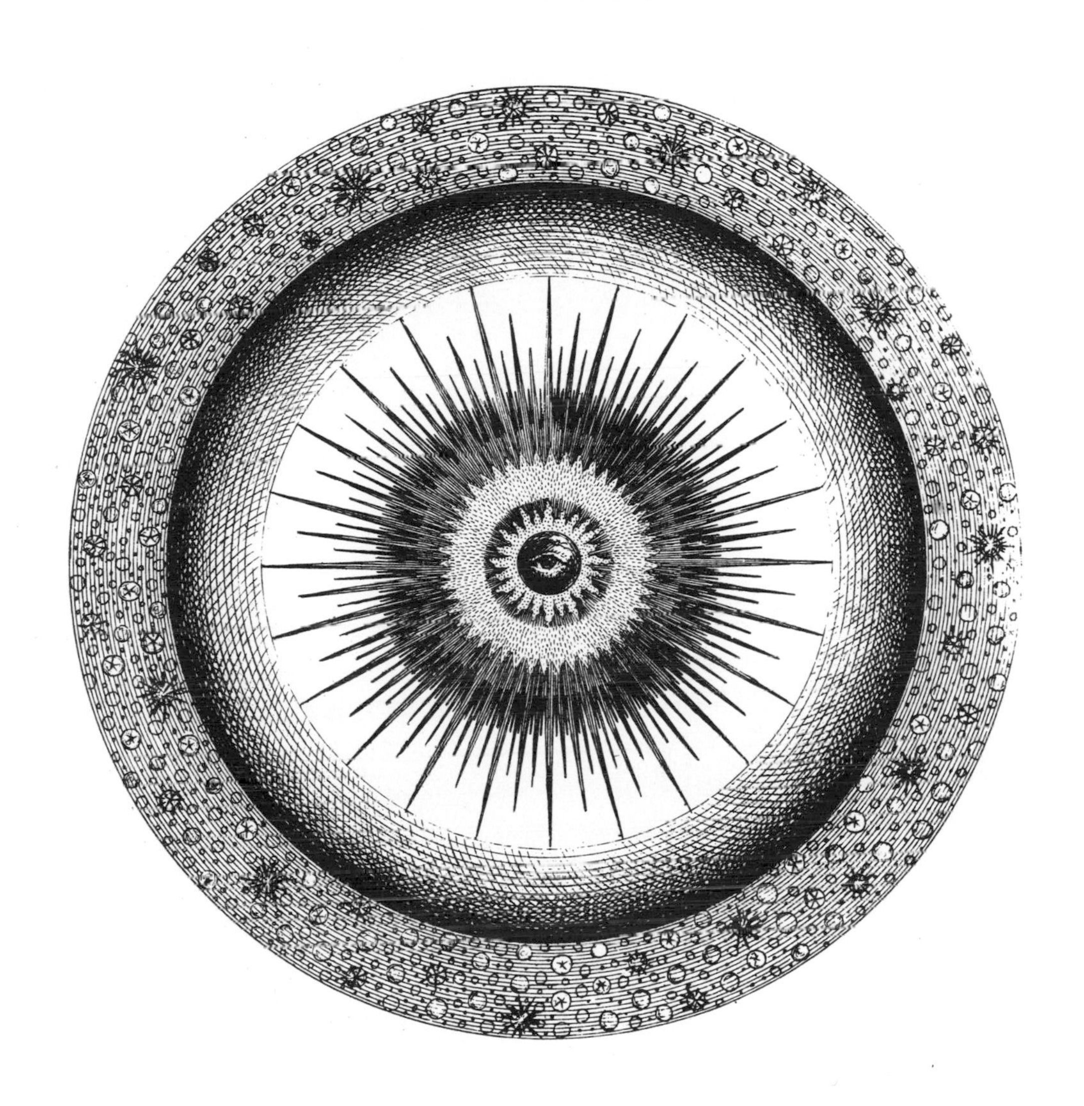

PLATE XXVI.

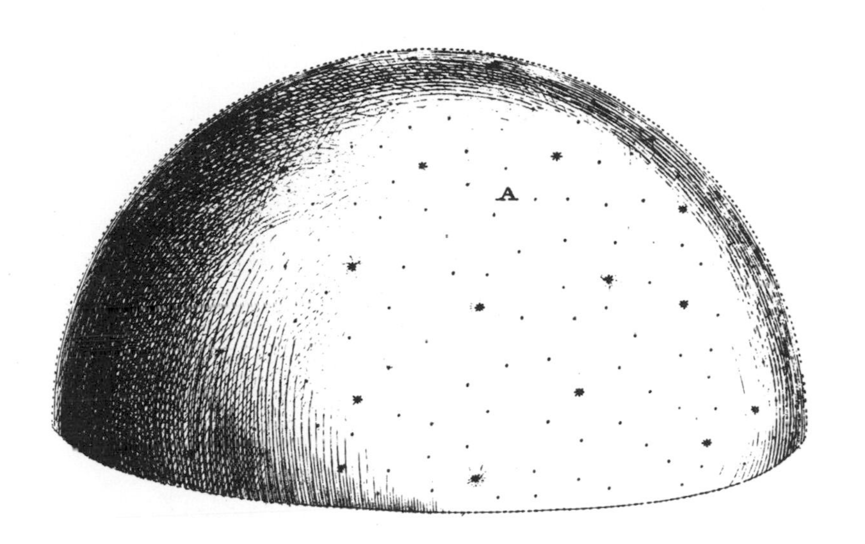

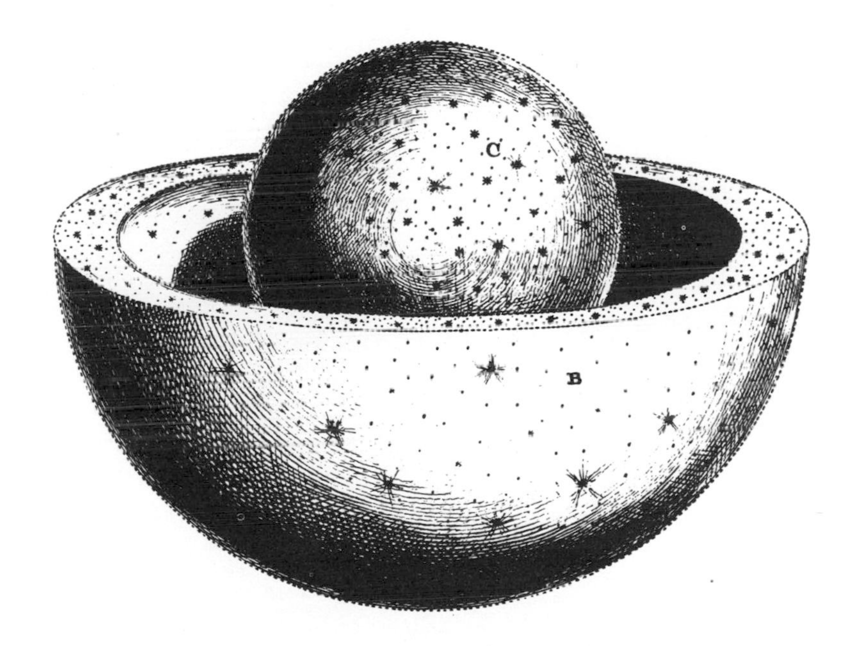

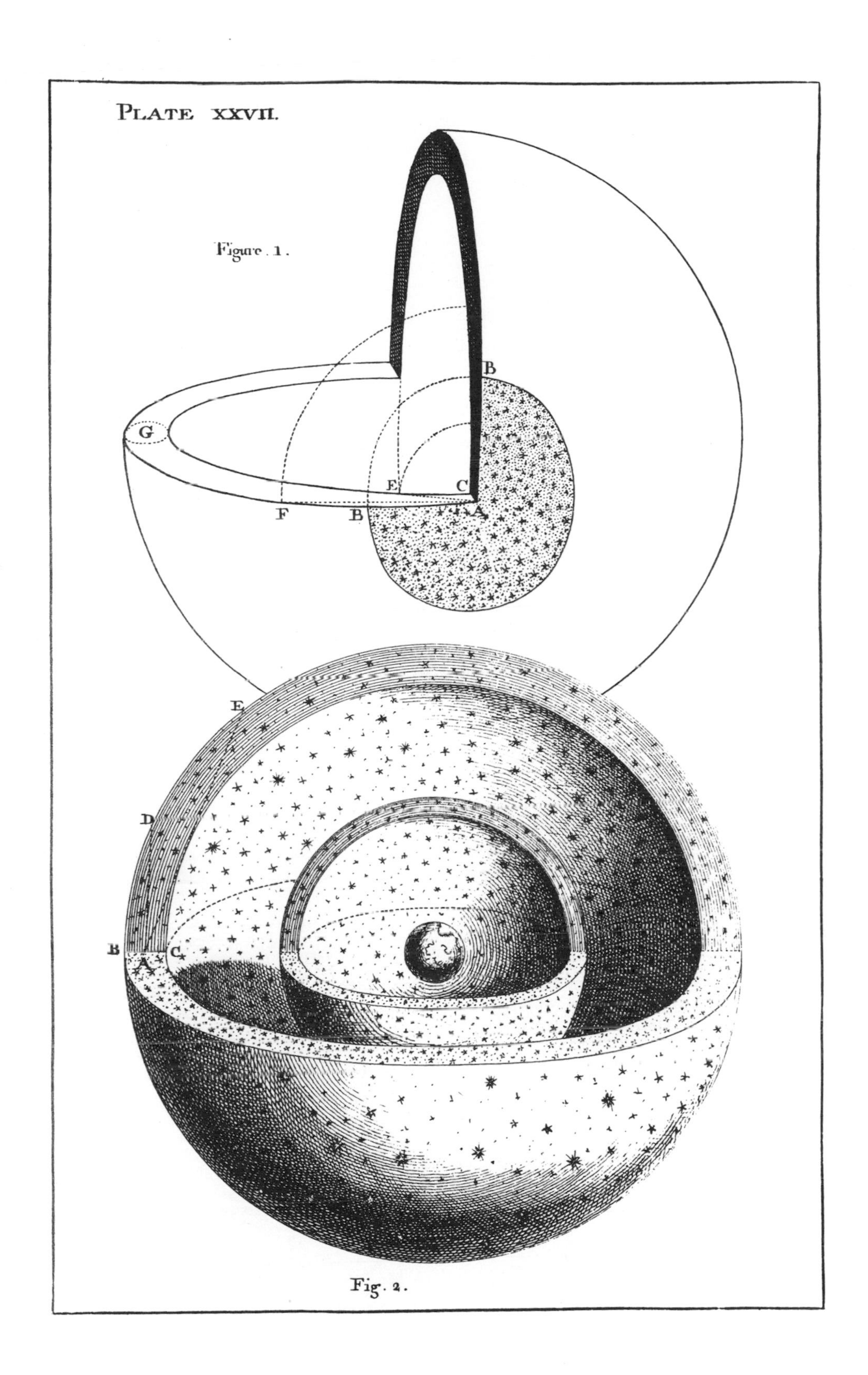
PLATE XXVII.
Figure. 1.
B
G
E
C
F
B
A
E
D
B
C
A
Fig. 2.

Figure I.

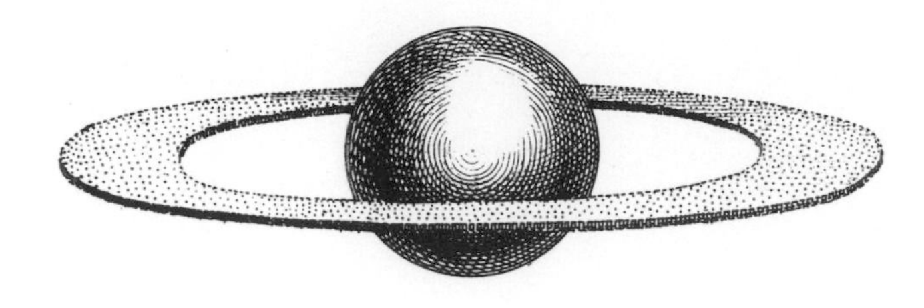

Fig. II.

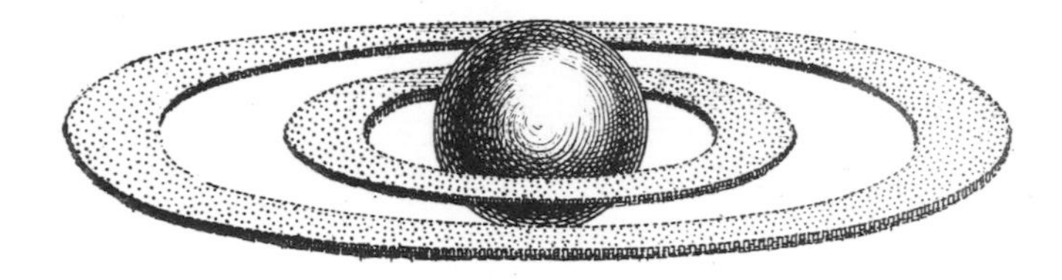

Fig. III.

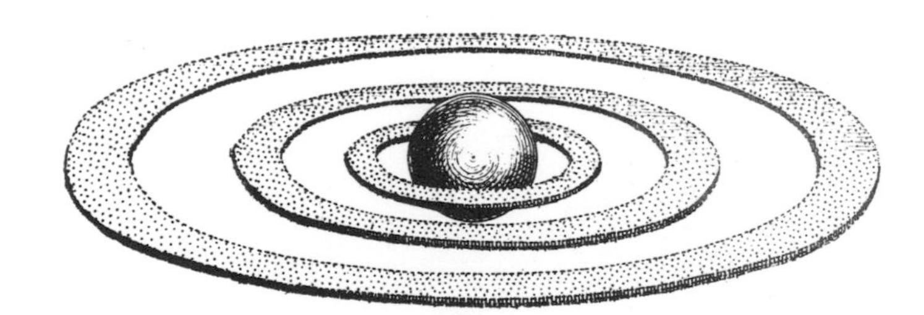

PLATE. XXIX.

Figure I.

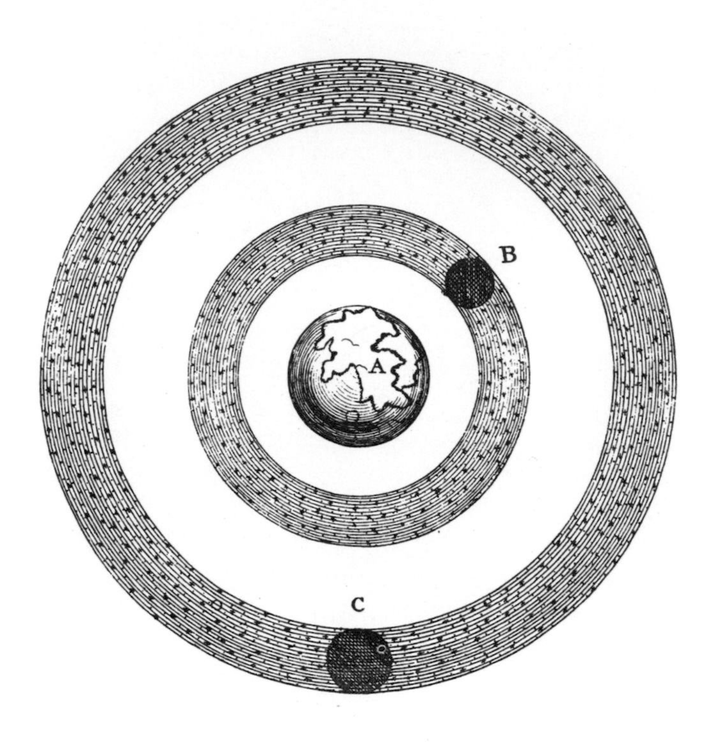

Fig. II.

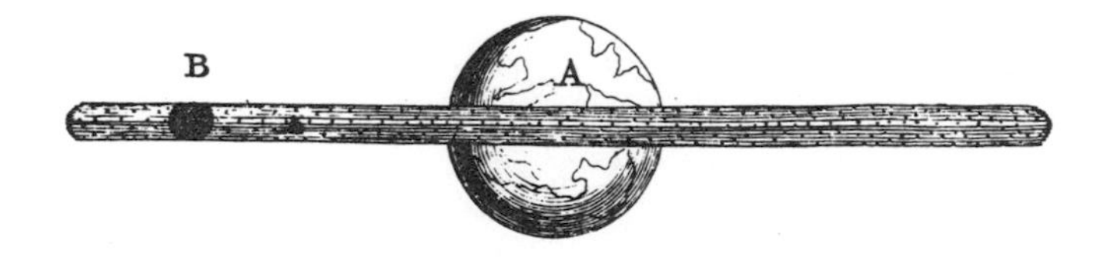

thus in a like regular Diſtribution of the Stars, there muſt of courſe be at leaſt three Times as many to be ſeen in this Direction of the Sine, or Semi-chord A E, itſelf, than in that of the ſemi-verſed Sine A C, or where near the Direction of the Radius of the Space G. *Q. E. D.*

But we are not confined by this Theory to this Form only, there may be various Syſtems of Stars, as well as of Planets, and differing probably as much in their Order and Diſtribution as the Zones of *Jupiter* do from the Rings of *Saturn*, it is not at all neceſſary, that every collective Body of Stars ſhould move in the ſame Direction, or after the ſame Model of Motion, but may as reaſonably be ſuppoſed as much to vary, as we find our Planets and Comets do.

Hence we may imagine ſome Creations of Stars may move in the Direction of perfect Spheres, all variouſly inclined, direct and retrograde; others again, as the primary Planets do, in a general Zone or Zodiack, or more properly in the Manner of *Saturn*'s Rings, nay, perhaps Ring within Ring, to a third or fourth Order, as ſhewn in *Plate* XXVIII. nothing being more evident, than that if all the Stars we ſee moved in one vaſt Ring, like thoſe of *Saturn*, round any central Body, or Point, the general Phænomena of our Stars would be ſolved by it; ſee *Plate* XXIX. *Fig.* 1. and 2. the one repreſenting a full Plane of theſe Motions, the other a Profile of them, and a viſible Creation at B and C, the central Body A, being ſuppoſed as *incognitum*, without the finite View; not only the Phænomena of the *Milky Way* may be thus accounted for, but alſo all the cloudy Spots, and irregular Diſtribution of them; and I cannot help being of Opinion, that could we view *Saturn* thro' a Teleſcope capable of it, we ſhould find his Rings no other than an infinite Number of leſſer Planets, inferior to thoſe we call his Satellites: What inclines me to believe it, is this, this Ring, or Collection of ſmall Bodies, appears to be ſometimes very excentric, that is, more diſtant from *Saturn*'s Body on one Side than on the other, and as viſibly leaving a larger Space between the Body and the Ring; which would hardly be the Caſe, if the Ring, or Rings, were connected, or ſolid, ſince we have good Reaſon to ſuppoſe, it would be equally attracted on all Sides by the Body of *Saturn*, and by that means preſerve every where an equal Diſtance from him; but if they are really little Planets, it is clearly demonſtrable from our own in like Caſes, that there may be frequently more of them on one Side, than on the other, and but very rarely, if ever, an equal Diſtribution of them all round the *Saturnian* Globe.

How much a Confirmation of this is to be wiſhed, your own Curioſity may make you judge, and here I leave it for the Opticians to determine. I ſhall content myſelf with obſerving that Nature never leaves us without a ſufficient Guide to conduct us through all the neceſſary Paths of Knowledge; and it is far from abſurd to ſuppoſe Providence may have every where throughout the whole Univerſe, interſperſed Modules of every Creation, as our Divines tell us, Man is the Image of God himſelf.

Thus, Sir, you have had my full Opinion, without the leaſt Reſerve, concerning the viſible Creation, conſidered as Part of the finite Univerſe; how far I have ſucceeded in my deſigned Solution of the *Via Lactea*, upon which the Theory of the Whole is formed, is a Thing will hardly be known in the preſent Century, as in all Probability it may require ſome Ages of Obſervation to diſcover the Truth of it.

It remains that I ſhould now give you ſome Idea of Time and Space; but this will afford Matter ſufficient for another Letter.

I am, &c.

LETTER

LETTER THE EIGHTH.

Of Time and Space, with regard to the known Objects of Immensity and Duration.

SIR,

THE Opportunity you gave me in your last Visit, of shewing you my general Scheme of the Universe, I find, besides the Pleasure it then gave, is now attended with many useful Advantages.

I now not only hope to be better understood for the future, but have reason to expect what I now write will merit your Attention more, and have some Title to your Approbation. The Ideas I have fram'd of Time and Space, will now more gradually fill your Imagination both with Wonder and Delight, before they can arise so high as to be lost in an Eternity and the Infinity of Space. And I am fully perswaded your farther Inquiries into these vast Properties of the Deity, will here be answered intirely to your Satisfaction. You must allow me now to be in some measure a Judge of what I think will please you most, from the Observations you have made upon my general System, or otherwise you would have reason to think me perhaps too presuming: But I flatter myself the great Difficulty is now over; and what remains to be said, will also naturally follow from what has gone before, that this Letter, I guess, will go near to furnish you with all the Ideas you wish to form upon the Subject. To what you have said of my having left out my own Habitation in my Scheme of the Universe, having travell'd so far into Infinity as both to lose sight of, and forget the Earth, I think I may justly answer as *Aristotle* did when *Alexander*, looking over a Map of the World, enquir'd of him for the City of *Macedon*; 'tis said the Philosopher told the Prince, That the Place he sought for was much too small to be there taken Notice of, and was not without sufficient Reason omitted.

The System of the Sun compar'd but with a very minute Part of the visible Creation, takes up so small a Portion of the known Universe, that in a very finite View of the Immensity of Space, I judg'd the Seat of the Earth to be of very little Consequence, could I have possibly represented it, as not only being one of the smallest Objects in our Regions, but in a manner infinitely leſs than even her own annual Orbit, and had nothing to do with my main Deſign, which was to repreſent all our planetary Worlds as one collective Body, and begin my comparative Scale of Magnitude from the Sun only and his Sphere of activity; as the ſmalleſt Object I could with any Propriety pretend to expreſs in ſuch a Plan.

In ſome Meaſure to convince you that I have committed no Error in this, I will try by ſome leſs mathematical Method than that of meer Numbers, to imprint an Idea in your Mind of the true Extent of the ſolar Syſtem, and the Magnitude of all its moving Bodies, by natural Objects moſt familiar to your Senſes. When we endeavour to form any Idea of Diſtance, Magnitude, or Duration, by Numbers only, we ſo ſoon exceed the Limits of Conception, that this way we find our Faculties of reaſoning as finite as our Senſes; and no doubt 'tis right it ſhould be ſo, Providence, as it were, having ordain'd that the firſt ſhould only attend the laſt, in ſuch an adequate Degree to a determin'd Diſtance; but what Diſtance or Degree of Knowledge is deſtin'd to human Nature, none but the Power that gave it can tell. 'Tis certain that beyond the third or fourth Place of our Nomenclator, we receive but very faint Impreſſions of the thing expreſt, and can frame ſcarce any Notion at all of either Number, Diſtance, or Magnitude, ſignified beyond it: Hence Aſtronomers are frequently oblig'd to have recourſe to mixt Ideas, and make Things of different Natures and Properties aſſiſt each other, to excite more adequate Ideas of what they would have conceived. Thus to expreſs immenſe Diſtances and Magnitude, they frequently apply themſelves to Time and Motion; and *vice verſa*, to ſignify a long Duration, they have often recourſe to Diſtance and Matter, removing, in Imagination, Worlds of Sand, Grain after Grain, to ſome remote known Region.

Heſiod, * to expreſs his Idea of the Diſtance from his higheſt Heaven to Earth, and from Earth to Hell, or *Tartarus*, ſuppoſes an Anvil to be let fall from one to the other, which he ſays in nine natural Days would reach the Earth from Heaven, and in the ſame time would fall from the Earth to Hell. † *Homer* makes his *Vulcan* fall from Heaven to the Iſland of *Lemnos* in much leſs Time, not exceeding one full artificial Day.

* From the high Heaven a brazen Anvil caſt,
Nine Nights and Days in rapid Whirls would laſt,
And reach the Earth the Tenth, whence ſtrongly hurl'd;
The ſame the Paſſage to th' infernal World.

COOKE.

† Hurl'd headlong downward from th' etherial Height;
Toſs'd all the Day in rapid Circles round,
Nor till the Sun deſcended touch'd the Ground.

POPE.

Modern

Modern Aſtronomers have made uſe of the ſwifteſt Velocity of a Cannon-Ball as continued thro' the Space they would ſo deſcribe, and in this Light, the Diſtance to the Sun has been by many compar'd to twenty-five Years Motion of a Cannon-Ball, ſuppoſing it to travel at the Rate of 100 Fathom in a Moment, *i. e. the Pulſe of an Artery*; and that a Journey ſo performed to one of the neareſt fix'd Stars, would take the ſame Body at leaſt 100,000 Years before it could arrive there. But the Method I have choſe to convey my Ideas of the Magnitude of the planetary Bodies, and the Extent of the viſible Creation to you, I am willing to hope you will find ſtill more familiar, comprehenſive, and eaſy: And it only depends upon your Remembrance of a very few known Objects, and their neighbouring Diſtances, which may be preſumed you are, or have been, very well acquainted with. You have not only very lately but very often been in *London*, and muſt, I think, retain ſome Idea of the Dome of St. *Paul*'s, tho' I own I ought not to be ſorry if you ſhould chance to have forgot it, provided it might prove a Means of making your Viſits more frequent. The Diameter of the Dome of this Church is 145 Feet: Now if you can imagine this to repreſent the Surface of the Sun, a ſpherical Body 18 Inches diameter, will juſtly repreſent the Earth in like Proportion; and another of only five Inches diameter, will repreſent the Moon. The Truths of theſe Proportions I have ſhewn in my *Clavis Cæleſtis*; and the Reaſon why I have here fixt upon the Dome of this Church for my firſt Object of Compariſon, will naturally appear from what follows.

From the Magnitude of the Earth on which we live, as from a known Scale with reſpect to its Parts compar'd with our own Bodies, we naturally frame our firſt Ideas of Extent, and fix our Rationale of Remoteneſs; by which we are ſufficiently enabled to judge of all other ſenſible Diſtances within one finite View. And hence by the undoubted Principles of Geometry, having firſt given the Meaſurement of the Earth in any known Proportion with any other Quantity moſt familiar to our Senſes, and the Angle of Appearance, or Parallax to any perceivable Object, we can eaſily find in homogenial Parts its true Diſtance from the Eye. And thus allowing for ſome ſmall tho' unavoidable Errors, that may poſſibly ariſe from the Difficulties of Obſervation (eſpecially ſmall Angles and minute Quantities) we can always determine to a ſufficient, and very frequently to a juſt Exactneſs, the relative Diſtance of all viſible Bodies, remote or near, ſuch as the Planets, Comets, and the Sun.

* In this Manner Aſtronomers having procûr'd a comparative Standard, reduc'd to ſome known Meaſure, as *Engliſh* Miles, Leagues, Semi-Orbs or Orbits,

* Parallax is the changeable Poſition of Bodies to different Situations of the Eye. Firſt having found the Quantity of a Degree (*i. e.* a 60th Part of the Circumference) upon the Earth's

Orbits, with all the Force of analogical Reaſoning, clearly can demonſtrate the Place and Diſtance of any Object within the Reach of Obſervation, and judge of Diſtances almoſt indefinite.

PLATE XXX.

Will help you to very correct Ideas of the real Magnitude of the Globe of the Earth, compar'd with the juſt Extent of the Iſland of *Great-Britain*, which you will find with *Ireland*, and the reſt of its Iſlands, ſeated near the Center of the Projection. This as a Standard will enable you to judge of all other Diſtances more perfectly; and firſt I ſhall conſider that of the Sun.

The Sun is found to be mean diſtant from the Earth nearly 81 Millions of Miles, or 6877,5 Diameters of the Earth; and *Saturn*, the remoteſt Planet from him is at his greateſt Diſtance from us about 858 Millions of Miles: Yet theſe Diſtances are but the beginning of Space, and only ſerve to open our Ideas for farther Search.

The great Comet of 1680, as I have ſome where ſaid before, was found to move in ſo vaſt an excentrick Orbit, that in its aphelion Point it would be 14,4 Times as far from the Sun, as the Orbit of *Saturn*, and hence at leaſt eleven thouſand and two hundred Millions of Miles from us. Now ſince the wiſe Creator hath ſo diſpos'd all the independent Parts of the Creation, ſuch as the ſeveral Syſtems of primary and ſecondary Planets, *&c.* at ſo great a Diſtance from each other, that the Laws of any one in no wiſe ſhall interfere, diſturb, or interrupt the Principles of another; this Comet, which we can eaſily prove belong'd to our own Sun, we may well imagine came not near any other; and tho' at that vaſt Diſtance from the ſolar Body, yet ſtill there muſt have remain'd a Space ſufficient to divide or ſeperate the ſenſible activity of neighbouring Syſtems, that they may not ruſh upon each other. Hence we may reaſonably ſuppoſe, that the neareſt Star can be no nearer than a triple Radius of its active Sphere; and provided they are all in regular Order, and much of the ſame Magnitude with one another (which no Arguments can poſſibly contradict) this Radius we may juſtly make 2000 times the Diſtance of our Earth. For admitting the utmoſt Limits of the Sun's Attraction to exceed this Sphere of the Comets, as far as the Sphere of the Comets

Earth's Surface, *Aratoſthenes* diſcover'd that the Magnitude of the whole was eaſily known; and then from the Moon's horizontal Parallax having given the Radius of the Earth, the Diſtance of the Moon is ſoon determined; next by the menſtrual Parallax of the Lunar Orbit, the Diſtance of the Sun is found; and by the Elongation of the inferior Planets, their mutual Diſtance from each other; and, laſtly, from the annual Parallax of the Earth's Orbit, all the other Orbits of the ſuperior Planets are eaſily found.

exceeds

THE

EARTH

PLATE XXX.

exceeds that of the Planets, which is nearly 14,4 times, the Radius of the ſolar Syſtem will be extended every way 200 Radius's of the Orbit of *Saturn*, and conſequently the Diſtance from Star to Star will not be leſs than 6000 times the Radius of our *Orbis Magnus*, and conſequently upwards of 480,000,000,000 Miles. That this is even leſs than the real Truth, and may be defended as a very moderate Computation, grounded upon Reaſon, we have infallible Demonſtration to witneſs, and make appear as thus.

We know from the Nature of Diſtance and Motion that the Stars may have an annual Parallax, but it is ſo very ſmall, that the very beſt Aſtronomers have never yet been able to aſſign what the Quantity really is. Yet it is allow'd by univerſal Conſent, that it can't poſſibly be more that one Minute of a Degree, and may probably be much leſs. Mr. *Flamſtead*, by repeated Obſervations, made it in ſome of them upwards of 40″; but Mr. *Bradley* has endeavour'd to prove it is every where too ſmall to be determined, and aſſigns this Angle to another Cauſe. This way then we cannot make their Diſtance leſs; and to prove that it is ſomething more than I have ſaid it is, let us even increaſe the doubtful Parallax of 40″ to the moſt it poſſibly can be, *viz.* to 60″ or 1′; and by the Solution of the Triangle, we ſhall find that the neareſt Star is 6875 times the Radius of the Earth's Orbit from the Sun: And this tho' more than any other Proportion makes them, is ſtill undeniably leſs than the Truth, which every Mathematician will of courſe be convinc'd of; and you yourſelf of force muſt believe, when you are told, that the ſmaller the Angle of Parallax is, the farther the Body is remov'd from us. By which Rule, according to Mr. *Flamſtead*'s Obſervations, the Diſtance muſt be ſtill greater: By the optical Experiment of * Mr. *Huygins*, greater ſtill than this; and according to Mr. *Bradley*, ſo much more as not even too be determin'd.

Now if the reſt are in general from each other, allowing the ſame Extent of Syſtem, and as much to part the like Extreams of active Virtue, be in ſuch Proportion of aerial Space, it will appear, that to paſs from any one Star to another, we muſt fly thro' ſo vaſt a Tract of pure Expanſe or Ether, that to viſit any one of the moſt neighbouring Syſtems, could we travel even as faſt as the ſwifteſt Eagle flies, for Inſtance, 500 Miles *per* Day, yet ſhould we be 3,000,000 of Years upon our way before we could arrive there; and if continuing on to view the Regions of the reſt within the known Creation, Myriads of Ages would be ſpent, and yet we could not hope to ſee the whole of but the ſmalleſt Conſtellation.

* 27664 Radius's of the *Orbis Magnus*, equal to the Diſtance of *Syrius*, whoſe Parallax ſhould be to anſwer it but 14″ 48‴.

But

But what Idea of Diſtance can you receive from this ſort of Eſtimation, where Numbers ariſe ſo very high. I own to you mine are ſoon quite loſt by this Method of counting, either, Diſtances or Duration. I believe few People can range their Ideas with ſuch Perſpicuity, as to arrive at any adequate Notion of any Number above a thouſand.

To give you therefore a clearer Idea of Diſtance, and impreſs the Proportions of Space more ſtrongly and fully in your Mind, let us ſuppoſe the Body of the Sun, as I have ſaid before, to be repreſented by the Dome of St. *Paul*'s; in ſuch Proportion a ſpherical Body eighteen Inches Diameter, moving at *Mary-le-bone*, will juſtly repreſent the Earth, and another of five Inches Diameter, deſcribing a Circle of forty-five Feet and a half Radius round it, will repreſent the Orbit and Globe of the Moon. A Body at the *Tower* of 9,7 Inches, will repreſent *Mercury*; and one of 17,9 Inches at St. *James*'s Palace will repreſent the Planet *Venus*; *Mars* may be ſuppoſed at a Diſtance, like that of *Kenſington* or *Greenwich*, 10 Inches Diameter: *Jupiter*, imagined to be at *Hampton-Court*, or *Dartford* in *Kent*; and *Saturn*, at *Cliefden*, or near *Chelmsford*: The firſt repreſented by a Globe 15 Foot 4 Inches Diameter, the latter by one of 11 Feet $\frac{1}{2}$ and his Ring four Feet broad: Theſe would all naturally repreſent the planetary Bodies of our Syſtem in their proper Orbits and proportional Magnitudes, as moving round the Cupola of St. *Paul*'s, as their common Center the Sun. And preſerving the ſame natural Scale, the Aphelion of the firſt Comet would be about *Bury*, the ſecond at *Briſtol*, and the third near the City of *Edinburgh*. But if you will take into your Idea one of the neareſt Stars; inſtead of the Dome of St. *Paul*'s, you muſt ſuppoſe the Sun to be repreſented by the gilt Ball upon the Top of it, and then will another ſuch upon the Top of St. *Peter*'s at *Rome* repreſent one of the neareſt Stars.

The whole Syſtem exhibited in the above Proportion, would be nearly as follows:

Diameter of the Sun	145	Feet.
Saturn	11,587,	his Ring 27,54, its Breadth 4.
Jupiter,	15,39.	
Mars,	10,15	Inches.
the Earth,	18,125.	
Venus,	17,98	
Mercury,	9,715	
and the Moon,	4,93	

Diſtance

* Distance of *Saturn* from the Sun,	27 Miles, and 1700 Yards.
Jupiter,	15 Miles, and 458 Yards.
Mars,	4 Miles, and 751 Yards.
the Earth,	2 Miles, and 1632 Yards.
Venus,	2 Miles, and 217 Yards.
Mercury,	1 Mile, and 267 Yards.

and of the Moon, from us, 45 Yards and a half.

That of the most distant Comet 390, and the nearest of the Stars not less than 6875, † Radius's of the *Orbis Magnus*.

Now, if like Creations crowd the vast Depths of Infinity, and if each are adapted to receive Beings of different Natures, where must our Wonder and Ideas have end?

As it is evident in the Sign *Taurus*, in *Perseus*, and *Orion*, that we can plainly perceive Stars to the sixth and ninth Magnitude, the former with our naked Eye, the other by the Help of Telescopes, the visional ocular Creation cannot be less than 4,320,000,000,000 Miles in semi Diameter, and admitting a regular Distribution of those primordial Bodies amongst themselves, the Depth, or most remote Limits of the *Vortex Magnus* from Side to Side, cannot be less than 8 m, m, 640 thousand of Million of Miles, admitting it is no more than what we see; and lastly, supposing our System to be situated nearly in the Middle of the *Vortex Magnus* (which, from the visible Order of the Stars, we may justly conjecture, with the highest Probability of Truth) the nearest Distance of the *Ens Primum*, in the Realms of eternal Day, will rise to 30,000,000,000,000 Miles, but more probably to 100,000,000,000,000 Miles, making the Confines of Creation from Verge to Verge in the first Case, upwards of 68 Million of Millions of Miles, Diameter, and by the last above 200'. But, if we compute the Distance of the Stars after the Manner of *Huygens*, for his Distance of *Syrius* from the Sun, the Distance of the Region of Immortality without exceeding Probability may rise to near 1,000,000,000,000,000 Miles.

Now to pass by any progressive Motion from the outward Verge, or Borders of the Creation, thro' the starry Regions of Mortality, if I may call

* Of the Satellites of *Saturn* in the above Proportion.

The		would be		
	1		27,96	Feet distant from his Center.
	2		35,52	
	3		50,	
	4		114,	
	5		341,9	

And those of *Jupiter*.

The		would be		
	1		28,51	Feet distant from him.
	2		69,177	
	3		110,224	
	4		190,	

	°	′	″			
* Radius, or Sign of	89	59	30	—	—	10,0000000
Sine substract of	0	0	30	—	—	6,1626961
Hence the Distance	6875,5			—	—	3,8373039

L

them

them ſo, as far as the Center of the *Ens Primum*, or *Sedes Beatorum*, according to *Homer*, or *Milton*'s Manner of meaſuring Space, a Body falling, or a Being moving with a Velocity but of 1000 Feet *per* Minute, *i. e.* at the Rate of 20,000 Yards *per* Hour, or about 300 Miles *per* Day, would be at leaſt 300,000,000 Years upon its Journey thither, if not 1,000, m, and perhaps much more, without offending Probability; but even three Million Centuries, or Ages, ſure is enough to be employ'd, in paſſing from one Place to another; therefore, we may conclude, the Soul muſt have ſome other Vehicle than can be found in the Ideas of Matter to convey it ſo far, at leaſt at once. Hence we may truly infer, that the Soul muſt be immaterial, and that in all Probability there may be States in the Univerſe ſo much more longer lived than ours, that, compared with the Age of Man, the Age of ſuch Beings may be almoſt as an Eternity, or rather, as that of the human Species to that of a Sun-born Inſect.

Again, if there are ſtill Stars beyond all theſe of other Denomination, which we do not here perceive, how vaſtly muſt theſe Numbers be increaſed, to expreſs, almoſt without Idea, the amazing Whole of this one viſible Creation; but what has been already ſaid, I judge will be ſufficient to ſhow the Immenſity of Space, and help you to conceive the ſtupendious Nature of an endleſs Univerſe; every where the home Poſſeſſion, Production, and inſtantaneous Care, of an infinite good Being, perfectly wiſe, and powerful, of whom we can have no Idea more, than a Being in dark Privation can have of Light, but through the Luſtre of his own reſplendent Attributes.

Thus, having attempted to enlarge your Ideas of the Creation in general, and in ſome meaſure having conſidered the Indefinity of Space, I ſhall in the next Place proceed to give you ſome Account of my Notions of Time.

As Diſtance is the Meaſure of Magnitude and of all Extent, and helps our Imagination to the Ideas of Space, ſo are progreſſive Moments the Meaſure of Velocity, and makes us ſenſible of Duration: And as Space may be extended through all Infinity, ſo Time may be continued as to Eternity. This Succeſſion of temporal Ideas impreſſed, or excited in the Mind, as an Effect of Matter in Motion, producing a perpetual Change, both of Objects earthly and celeſtial, enables us not only to reflect upon paſt Viciſſitudes of Nature, but from their regular Courſes, known Order and Returns, predict Phænomena to come, and prove the periodical Effects of Nature's conſtant Laws ſo juſt and certain, that Time may be ſaid with Truth, to co-exiſt with Motion.

Meaſure being a certain Quantity of Senſation interwove with our Ideas of Diſtance and Duration, proceeding from a Reflection of what is impreſſed upon the Mind by ſome external Object, I muſt again return to our Mother of Ideas the Earth, and from thence, as I did, of Diſtance, frame

frame the original Images beſt ſuited to the Underſtanding, proper for our Judgment of Duration.

Time takes its firſt Denomination from the diurnal Rotation of the Earth upon its Axis, which we call a natural Day, and this for obvious Reaſons we ſubdivide in twenty-four Parts or Hours. This diurnal Motion having been ſucceſſively repeated, and the Day renewed three hundred and ſixty-five Times, we find that all the vegetable World has gone through all its Variegations, and Nature has again put on the ſame Face, adapted to the Seaſon; during which Time, and indeed which occaſions this general Change and Repetition, the Earth is found to make one intire Revolution round the Sun. This Space, or Period of Time, we call a ſolar, or rather a natural Year; and from our Senſibility of this, and its conſtituent Parts, both horary and diurnal, we form our general Judgment of Duration.

Saturn, the moſt remote, and moſt regular Planet in our Syſtem, as has been ſaid before, performs one Revolution round the Sun in about twenty-nine of the above ſolar Years: The great Comet of 1680 makes but one periodical Return in five hundred and ſeventy-five of thoſe Years, and the general Motion of the Stars, ariſing from the Proceſſion of the Equinoxes, altogether continually changing their Aſpect, or Poſition, at the Rate of 50″ *per* Year round the ecliptic Poles, compleats but one Revolution in 25920 Years; in which Time the whole ſidereal Frame of Heaven has changed, and every Star returned to the ſame Point of the ſolar Sphere it ſet out from. This is by many called the great *Saturnian* Year: Concerning which, Mr. *Addiſon* has thus tranſlated an eminent Author.

> When round the great *Saturnian* Year has turn'd,
> In their old Ranks the wandering Stars ſhall ſtand,
> As when firſt marſhall'd by the Almighty's Hand.
>
> ADDISON.

Now, if this ſidereal Revolution, ariſing from a ſecondary Cauſe, require this Number of Years to perfect one Rotation, what muſt their primitive Orbits take to circumſcribe the *Vortex Magnus*.

It has been obſerved, that the biggeſt Star to us ſcarce moves a Minute in an hundred Years, and the moſt remote as inſenſibly for Ages, from whence and what has been already ſaid of the imagined Diſtance of the general Center, we may frame this probable and well-grounded Gueſs, that the mean Revolution of a Star near the Middle of the *Vortex Magnus*, cannot be made in leſs than a Million of Years, and though to us imperceptible, our Sun in his own orbicular Direction, may be moving many Miles *per* Day. Beſides, if local Motion can be proved amongſt the Stars, what leſs than an Eternity can again reſtore them to their original Order and primitive State.

 Such

Such vaſt Room in Nature, as *Milton* finely expreſſes it, cannot be without its Uſe; and nothing but abſolute Demonſtration is wanting (which from their Nature and Diſtance cannot be expected) to confirm the grand Deſign, ſo ſuited to the Deity's infinite Capacity, and of eternal Benefit to all his Creatures, eſpecially Beings of a rational Senſe, and in particular Mankind.

Of theſe habitable Worlds, ſuch as the Earth, all which we may ſuppoſe to be alſo of a terreſtrial or terraqueous Nature, and filled with Beings of the human Species, ſubject to Mortality, it may not be amiſs in this Place to compute how many may be conceived within our finite View every clear Star-light Night. It has already been made appear, that there cannot poſſibly be leſs than 10,000,000 Suns, or Stars, within the Radius of the viſible Creation; and admitting them all to have each but an equal Number of primary Planets moving round them, it follows that there muſt be within the whole celeſtial Area 60,000,000 planetary Worlds like ours. And if to theſe we add thoſe of the ſecondary Claſs, ſuch as the Moon, which we may naturally ſuppoſe to attend particular primary ones, and every Syſtem more or leſs of them as well as here; ſuch Satellites may amount in the Whole perhaps to 100,000,000, or more, in all together then we may ſafely reckon 170,000,000, and yet be much within Compaſs, excluſive of the Comets which I judge to be by far the moſt numerous Part of the Creation.

In this great Celeſtial Creation, the Cataſtrophy of a World, ſuch as ours, or even the total Diſſolution of a Syſtem of Worlds, may poſſibly be no more to the great Author of Nature, than the moſt common Accident in Life with us, and in all Probability ſuch final and general Doom-Days may be as frequent there, as even Birth-Days, or Mortality with us upon the Earth.

This Idea has ſomething ſo chearful in it, that I own I can never look upon the Stars without wondering why the whole World does not become Aſtronomers; and that Men endowed with Senſe and Reaſon, ſhould neglect a Science they are naturally ſo much intereſted in, and ſo capable of inlarging the Underſtanding, as next to a Demonſtration, muſt convince them of their Immortality, and reconcile them to all thoſe little Difficulties incident to human Nature, without the leaſt Anxiety.

Such a Protheſis can ſcarce be called leſs than an ocular Revelation, not only ſhewing us how reaſonable it is to expect a future Life, but as it were, pointing out to us the Buſineſs of an Eternity, and what we may with the greateſt Confidence expect from the eternal Providence, dignifying our Natures with ſomething analogous to the Knowledge we attribute to Angels; from whence we ought to deſpiſe all the Viciſſitudes of adverſe Fortune, which make ſo many narrow-minded Mortals miſerable.

I am now, &c.

LETTER THE NINTH.

Reflections, by Way of General Scolia, *of Consequences relating to the Immortality of the Soul, and concerning Infinity and Eternity.*

S I R,

THIS my last Letter to you, I mean my final astronomical one, I propose as a *General Scolia* to the rest, the principle Matter being Reflections upon what is gone before, with some Conclusion naturally following or appendant to what has been already said; but which, I could not in any other Place, so properly remark to you.

The Probability of the foregoing Conjectures, chiefly built upon very distant Observations, shew an apparent Necessity for some other kind of Doctrine permitted by Providence, to give Mankind a Knowledge of their Immortality and Dependance upon it, in the first Ages of the World.

And for the same Reason it evidently appears, that the ancient Philosophers had it not in their Power to prove a supream *Being* and Director of all Things this Way.

And yet, as by a Sort of Instinct, or natural Reason, and Consciousness of a *good Principle,* we see how many noble Steps they made towards it, and was convinc'd at last of this *great Truth,* that since there was a *Mind* in so imperfect a Creature as Man, the *perfect Universe,* which comprehended all Things, could not possibly be without one; and as Sir *Isaac Newton* has justly observed in his *Principia,* " If every Par-" ticle of Space be *always,* and every individual Moment of Duration " *every where*; surely the Maker and Lord of all Things, cannot be *never* " and *no where.*"

To make manifest the infinite Empire and Agency of God, from celestial Motion, became the Task, but of very late Years; and I can't help being of Opinion, that by means of these primary Bodies, only, we shall at length be able to trace the greater Circulations, and Laws of Nature, to their real original and fountain Head.

These

These, were any thing wanting, besides the *Miracle ourselves*, to convince us of a divine Origination, are all infallible Proofs, that the Universe is governed by an intelligent and all-powerful Being, whose Existence is too nearly related to a self-evident Truth to be more clearly demonstated, than it is manifest of itself, both from the particular Laws of Nature, and the general Order of Things. An Argument which has been thought of no small Force, and well worth observing in the Infancy of *Christianity*. *The invisible Things of God are clearly seen, being understood by the Things that are made, even his eternal Power and Godhead.* Rom. i. 20.

But 'tis now high time to look back upon my Theory, and tell you it is a vain Supposition, to imagine I shall ever be able to convince every Reader, either of the Truth or Probability of what I have advanced to you: Mathematical Assistance not being to be expected, where perhaps it has never been thought of; and I allow you, it is much more reasonable to expect, that fifty Persons will read these Letters without perceiving the Reasonableness of them, than that five should consider them with proper Judgment.

I must ingenuously confess to you, that nothing is wanting to convince me intirely of the Certainty of what I here advance by way of Conjecture to you. But this you must only look upon as an happy Partiality, which generally attends all Authors, and always will be the chief Support of their tedious Labours. I assure you, I have neither Hopes nor Expectation, no, not the weak Breath of a Wish, to be admitted a proper Judge of my own Works. But I shall always take their Imperfection to be rather, (like my own Faults) to be too near me to be seen; I therefore trust all to my Friend, and if I am so fortunate as to excite his Approbation, I shall think myself very happy in a very favourite Point; which is, The advancing nothing which a rational Reader would willingly overlook, or be ignorant of.

But if I have been so happy as to come so near the Mark, as to border upon Truth, I believe you will allow me to carry my Conjectures a little further, and point out some farther pleasing Consequences, which I begin to perceive may naturally follow.

Should it be granted, that the Creation may be circular or orbicular, I would next suppose, in the general Center of the whole an intelligent Principle, from whence proceeds that mystick and paternal Power, productive of all Life, Light, and the Infinity of Things.

Here the to-all extending Eye of Providence, within the Sphere of its Activity, and as omnipresently presiding, seated in the Center of Infinity, I would imagine views all the Objects of his Power at once, and every Thing immediately direct, dispensing instantaneously its enlivening Influence, to the remoteſt Regions every where all round. I know you'll ſay Aſtronomers are never to be ſatisfied, and I muſt own where there is ſo much rational Entertainment for the human Mind, and ſo ſuitable to the true Dignity of God, and moſt worthy of Man, it is not eaſy to know where to ſtop in ſuch a Scene of Wonders.

Having, I ſay, once granted that all the Stars may move round one common Center, I think it is very natural to one, who loves to purſue Nature as far as we may, to enquire what moſt likely may be in that Center; for ſince we muſt allow it to be far ſuperior to any other Point of Situation in the known Univerſe, it is highly probable, there may be ſome one Body of ſiderial or earthy Subſtance ſeated there, where the divine Preſence, or ſome corporeal Agent, full of all Virtues and Perfections, more immediately preſides over his own Creation. And here this primary Agent of the omnipotent and eternal Being, may ſit enthroned, as in the *Primum Mobile* of Nature, acting in Concert with the eternal Will. To this common Center of Gravitation, which may be ſuppoſed to attract all Vertues, and repel all Vice, all Beings as to Perfection may tend; and from hence all Bodies firſt derive their Spring of Action, and are directed in their various Motions.

Thus in the *Focus*, or Center of Creation, I would willingly introduce a primitive Fountain, perpetually overflowing with divine Grace, from whence all the Laws of Nature have their Origin, and this I think would reduce the whole Univerſe into regular Order and juſt Harmony, and at the ſame time, inlarge our Ideas of the divine Indulgence, open our Proſpect into Nature's fair Vineyard, the vaſt Field of all our future Inheritance.

But what this central Body really is, I ſhall not here preſume to ſay, yet I can't help obſerving it muſt of Neceſſity, if the Creation is real and not merely Ideal, be either a Globe of Fire ſuperior to the Sun, or otherwiſe a vaſt terraqueous or terreſtial Sphere, ſurrounded with an Æther like our Earth, but more refined, tranſparent and ſerene. Which of theſe is moſt probable, I ſhall leave undetermined, and muſt acknowledge at the ſame time, my Notions here are ſo imperfect, I hardly dare conjecture. 'Tis true, I have ventur'd to think it may be one of theſe, and ſince ſo glorious a Situation can hardly be ſuppoſed without its proper Inhabitants, 'tis moſt natural to conclude it may be the latter. In the firſt Caſe, beſides our having no Idea of Beings exiſting in Fire, it would not, notwithſtanding its Diſtance, be ſo eaſy to account for its being inviſible; and ſince the Luſtre of the Stars are all innate, they could receive no Benefit from it, and conſequently ſuch a Nature as a ſolar Compoſition, muſt in this Place be render'd uſeleſs; but in the latter Sup-

 poſition.

position of its being a dark Body, we have no Difficulty attending us, having several Instances of like Bodies, moving round an opaque one. Now when we consider, that all those radient Globes, which adorn the Skies, those bright ætherial Sparks of elemental Fire, thick strewed like Seeds of Light all round our Hemisphere, are each to us the Embrio of a glorious Sun ; how awful and stupendious must that Region be, where all their Beams unite and make one inconceivable eternal Day ?

Though the Deity, says a learned Writer " be essentially present thro' " all the Immensity of Space, there is one Part of it in which he disco- " vers himself in a most transcendent and visible Glory. This is that Place " which is mark'd out in Scripture, under the different Appellations " of PARADICE ; *the third Heaven* ; *the Throne of* GOD, *and the Habi-* " *tation of his Glory*."

THIS continues the same Author, is " that Presence of God, which " some of the Divines call his glorious, and others his majestick Pre- " sence."

It is here, and here only, as in the Center of his infinite Creations, where he resides in a sensible Magnificence, and in the midst of those Splendors, which can Effect the Imagination of his Creatures ; and though the most sacred and supreme Divinity be allowed as essentially present in all other Places as well as in this, as being a BEING whose Center is every where, and Circumference no where ; yet it is here only, or in such Sensorium of his Unity, where he manifests his corporeal Agency, as in the Foci of his infinite Empire over all created Beings. It is to this majestick Presence of GOD, we may apply those beautiful Expressions of Scripture, " *Behold even to the Moon and it shineth not* ; *yea the Stars* " *are not pure in his Sight*."

" The Light of the Sun, and all the Glories of the World, on which " we live, are but as weak and sickly Glimmerings, or rather Dark- " ness it self, in Comparison of those Splendors, which encompass this " Throne of GOD."

Here Heav'ns wide Realms an endless Scene displays,
And Floods of Glory thro' its Portals blaze ;
The Sun himself lost in superior Light,
No more renews the Day, or drives away the Night :
The Moon, the Stars, and Planets disappear,
And Nature fix't makes one eternal Year.

Here and here alone center'd in the Realms of inexpressible Glory, we justly may imagine that primogenial Globe or Sphere of all Perfections, subject

ſubject to the Extreams of neither Cold nor Heat, of eternal Temperance and Duration. Here we may not irrationally ſuppoſe the Vertues of the meritorious are at laſt rewarded and received into the full Poſſeſſion of every Happineſs, and to perfect Joy. The final and immortal State ordain'd for ſuch human Beings, as have paſſed this Vortex of Probation thro' all the Degrees of human Nature with the ſupream Applauſe.

What vaſt room is here, for infinite Power and Wiſdom to act in, and that ſo viſibly takes Delight to bleſs all his Beings with his Bounty. And endleſs as his Preſcience, Attributes, and Goodneſs, are undoubtedly all thoſe natural and apparent Joys with which he manifeſts his Love to all his Creatures, a Multiplicity of Objects not to be enumerated. For whereſoever we turn our Eyes, and follow with our Reaſon, we may meet with Worlds of all Formations, ſuited no doubt to all Natures, Taſtes, and Tempers, and every Claſs of Beings.

Here a Groupe of Worlds, all Vallies, Lakes, and Rivers, adorn'd with Mountains, Woods, and Lawns, Caſcades and natural Fountains; there Worlds all fertile Iſlands, cover'd with Woods, perhaps upon a common Sea, and fill'd with Grottoes and romantick Caves. This Way, Worlds all Earth, with vaſt extenſive Lawns and Viſtoes, bounded with perpetual Greens, and interſperſed with Groves and Wilderneſſes, full of all Varieties of Fruits and Flowers. That World ſubſiſting perhaps by ſoft Rains, this by daily Dews, and Vapours; and a third by a central, ſubtle Moiſture, ariſing like an Effluvia, through the Pores and Veins of the Earth, and exhaling or abſorbing as the Seaſon varies to anſwer Nature's Calls. Round ſome perhaps, ſo denſe an Atmoſphere, that the Inhabitants may fly from Place to Place, or be drawn through the Air in winged Chariots, and even ſleep upon the Waves with Safety; round others poſſibly, ſo thin a fluid, that the Arts of Navigation may be totally unknown to it, and look'd upon as impracticable and abſurd, as Chariot flying may be here with us; and ſome where not improbably, ſuperior Beings to the human, may reſide, and Man may be of a very inferior Claſs; the ſecond, third, or fourth perhaps, and ſcarce allow'd to be a rational Creature. Worlds, with various Moons we know of already; Worlds, with Stars and Comets only, we equally can prove is very probable; and that there may be Worlds with various Suns, is not impoſſible. And hence it is obvious, that there may not be a Scene of Joy, which Poetry can paint, or Religion promiſe; but ſomewhere in the Univerſe it is prepared for the meritorious Part of Mankind. Thus all Infinity is full of States of Bliſs; Angelic Choirs, Regions of Heroes, and Realms of Demi-Gods; Elyſian Fields, Pindaric Shades, and Myriads of inchanting Manſions,

M. not

not to be conceived either by Philoſophy or Fancy, aſſiſted by the ſtrongeſt Genius and warmeſt Imagination.

All harmoniouſly crowded and provided with every Object of Beatitude, that Friendſhip, Love, or Society can inſpire, the Muſes or the Graces Frame; and all as permanent and perfect, that is deſtin'd to a Duration, ſuited to the Nature of their Exiſtence and Degree of Cogniſance; for as a very learned Writer obſerves upon this ſame Subject;

" How can we tell, but that there may be above us Beings of greater
" Powers, and more perfect Intellects, and capable of mighty Things,
" which yet may have corporeal Vehicles as we have, but *finer* and
" *inviſible*? Nay, who knows, but that there may be even of theſe
" many *Orders*, riſing in Dignity of Nature, and Amplitude of Power,
" one above another? It is no Way below the Philoſophy of theſe Times,
" which ſeems to delight in inlarging the Capacities of Matter, to aſſert
" the Poſſibility of this."

From theſe amazing Ideas of Space in general, and from the particular Diſtance of the Stars, which ſeparates as it were, one Syſtem of Bodies from another, and by ſo prodigious an extent, as ſcarce to be ſuppos'd a temporal Taſk. I think it naturally follows, had we no other Way to prove it, or any other Reaſon to believe it, that the Soul muſt of Neceſſity be immaterial; for as this Space ſeems ſo impaſſible to Matter, as not to be undertaken and performed without the Loſs of Ages, in a State only of Tranſmigration, we may well imagine, that Change of Place is not effected this Way, but by ſome other Vertue or Property, more immediate, if not inſtantaneous.

I own next to *Annihilation* is the State of Oblivion, and this Way we may ſolve all Difficulties with regard to our being ſenſible of ſuch a Loſs of Exiſtence; but if we allow the Soul to be immaterial, it no longer has any thing to do with Space, but as operating by Reflection only, or the Faculty of Thinking; it may be like the Imagination where it pleaſes in a Moment.

Objects of the Mind abſtracted from the Senſes of the Body, has no real or comparative Magnitude; that is, I would ſay, an Inch, a Foot, a Yard, a Mile, or a Million of Miles are all equally indefinite, and is thus prov'd; every finite Line is formed of an infinite Number of Points, and no finite Line can be ſolv'd into more. Thus if you will allow me the Expreſſion, the Mind being magnified as all Objects are diminiſhed, what ſeems impracticable in the natural State of Things, in an Ideal one, becomes very poſſible; that is, to make myſelf more intelligible, though we can hardly conceive, how any Being can paſs from *Syrius* to the Sun, by natural Laws in their proper State, yet if proportionally reduced by a

new

PLATE. XXXI.

PLATE XXXII.
A
H
F
B
D
I
E
C
G

new Modification of Ideas, to the Bigneſs of a Ball 6 Feet Diameter, and to be only 680 Miles aſunder; the Thing is very comprehenſive and eaſy.

Hence Viſion, Light, and Electrical Virtue, ſeem to be propagated with ſuch Velocity, that nothing but God can poſſible be the Vehicle; and hence we may juſtly ſay with St. *Paul*, *Acts* xvii. 28. *In him we live, in him we move, in him we have our Being.*

It will further appear, from the foregoing Letters, that all the Stars and planetary Bodies within the finite View, are altogether but a very minute Part of the whole rational Creation; I mean that vaſt collective Body of habitable Beings, which I have endeavoured to demonſtrate, are all govern'd by the ſame Laws, though variouſly revolving round one common Center, in which Center we may not impertinently venture to ſuppoſe the prime Agent of our Natures; or otherwiſe, the moſt perfect of all created Beings, illimitable in his Ideas and Faculties of Senſation particularly preſide.

> But tho' paſt all diffus'd, without a Shore
> His Eſſence; *local* is his Throne, (as meet)
> To gather the diſperſt, (as Standards call
> The liſted from afar) to fix a Point;
> A central Point, collective of his Suns,
> Since finite ev'ry Nature, but his own. Dr. *Young*.

And farther ſince without any Impiety; ſince as the Creation is, ſo is the Creator alſo magnified, we may conclude in Conſequence of an Infinity, and an infinite all-active Power; that as the viſible Creation is ſuppoſed to be full of ſiderial Syſtems and planetary Worlds, ſo on, in like ſimilar Manner, the endleſs Immenſity is an unlimited Plenum of Creations not unlike the known Univerſe. See *Plate* XXXI. which you may if you pleaſe, call a partial View of Immenſity, or without much Impropriety perhaps, a finite View of Infinity, and all theſe together, probably diverſified; as at A, B and C. in *Plate* XXXII. which repreſents their Sections, if all may be a proper Term for an infinite or indefinite Number, we may juſtly imagine to be the Object of that incomprehenſible Being, which alone and in himſelf comprehends and conſtitutes ſupreme Perfection.

That this in all Probability may be the real Caſe, is in ſome Degree made evident by the many cloudy Spots, juſt perceivable by us, as far without our ſtarry Regions, in which tho' viſibly luminous Spaces, no one Star or particular conſtituent Body can poſſibly be diſtinguiſhed; thoſe in all

all likelyhood may be external Creation, bordering upon the known one, too remote for even our Telescopes to reach.

With the raptur'd Poet may we not justly say

> O, what a Root! O what a Branch is here!
> O what a Father! what a Family!
> Worlds! Systems! and Creations!

And in Consequence of this

> In an Eternity, what Scenes shall strike?
> Adventures thicken? Novelties surprize?
> What Webs of Wonder shall unravel there?
>
> *Night Thoughts.*

So many varied Seats where every Element may have its proper Beings and all adapted to partake of every thing suited to their Natures, argue such Maturity of Wisdom, and the vast Production such mysterious Power; 'tis hardly possible for Mortals not to see divine Intelligence preside, and that every Being somewhere must be happy.

A Universe so well designed, and fill'd with such an endless Structure of material Beings, and all the Result of Prescience and infinite reflected Reason, flowing from a Mind all perfect, full of all Ideas, could never be designed in vain; and tho' our narrow Bounds of Reason limited, by finite Senses, cannot directly see the Consequence dependant on a Sequel, yet from what we do see, great Room we have to hope the next Stage of Existence will be more lasting and more perfect; and it is highly probable, the noblest Suggestion of the most luxuriant Fancy may fall infinitely short of what we are designed for.

But here, even in this World, are Joys which our Ideas of Heaven can scarce exceed, and if Imperfection appear thus lovely, what must Perfection be, and what may we not expect and hope for, by a meritorious Acquiescence in Providence, under the Direction, Indulgence, and Protection of infinite Wisdom and Goodness, who manifestly designs perfect Felicity, as the Reward of Virtue in all his Creatures, and will at proper Periods answer all our Wishes in some predestined World.

All this the vast apparent Provision in the starry Mansions, seem to promise: What ought we then not to do, to preserve our natural Birthright to it and to merit such Inheritance, which alas we think created all to gratify alone, a Race of vain-glorious gigantick Beings, while they are confined to this World, chained like so many Atoms to a Grain of Sand.

I am, &c.